From Molecules to Medicines

NATO Science for Peace and Security Series

This Series presents the results of scientific meetings supported under the NATO Programme: Science for Peace and Security (SPS).

The NATO SPS Programme supports meetings in the following Key Priority areas: (1) Defence Against Terrorism; (2) Countering other Threats to Security and (3) NATO, Partner and Mediterranean Dialogue Country Priorities. The types of meeting supported are generally "Advanced Study Institutes" and "Advanced Research Workshops". The NATO SPS Series collects together the results of these meetings. The meetings are co-organized by scientists from NATO countries and scientists from NATO's "Partner" or "Mediterranean Dialogue" countries. The observations and recommendations made at the meetings, as well as the contents of the volumes in the Series, reflect those of participants and contributors only; they should not necessarily be regarded as reflecting NATO views or policy.

Advanced Study Institutes (ASI) are high-level tutorial courses intended to convey the latest developments in a subject to an advanced-level audience

Advanced Research Workshops (ARW) are expert meetings where an intense but informal exchange of views at the frontiers of a subject aims at identifying directions for future action

Following a transformation of the programme in 2006 the Series has been re-named and re-organised. Recent volumes on topics not related to security, which result from meetings supported under the programme earlier, may be found in the NATO Science Series.

The Series is published by IOS Press, Amsterdam, and Springer, Dordrecht, in conjunction with the NATO Public Diplomacy Division.

Sub-Series

A.	Chemistry and Biology	Springer
B.	Physics and Biophysics	Springer
C.	Environmental Security	Springer
D.	Information and Communication Security	IOS Press
E.	Human and Societal Dynamics	IOS Press

http://www.nato.int/science
http://www.springer.com
http://www.iospress.nl

Series A: Chemistry and Biology

From Molecules to Medicines

Structure of Biological Macromolecules and Its Relevance in Combating New Diseases and Bioterrorism

edited by

Joel L. Sussman
Department of Structural Biology
Weizmann Institute of Science
Rehovot, Israel

and

Paola Spadon
Department of Chemical Sciences
University of Padova, Italy

Springer

Published in cooperation with NATO Public Diplomacy Division

Proceedings of the NATO Advanced Study Institute on
From Molecules to Medicines: Integrating Crystallography in the Fight against Bioterrorism and Emerging Diseases affecting Security
Erice, Italy
29 May – 8 June 2008

Library of Congress Control Number: 2009926163

ISBN 978-90-481-2338-4
ISBN 978-90-481-2337-7
ISBN 978-90-481-2339-1 (eBook)

Published by Springer,
P.O. Box 17, 3300 AA Dordrecht, The Netherlands.

www.springer.com

Printed on acid-free paper

CONTENTS

PREFACE

This volume comprises papers presented at the 40th Erice Course "From Molecules to Medicine: Structure of Biological Macromolecules and Its Relevance in Combating New Diseases and Bioterrorism", May 29 to June 8, 2008. The papers span the breadth of material presented, which emphasize the practical aspects of modern macromolecular crystallography and its applications to medicine.

Topics addressed span from the selection of targets, through to structure determination, interpretation and exploitation. A particular theme that emerges is the dependence of modern structural science on multiple experimental and computational techniques. It is both the development of these techniques and their integration that will take us forward in the future.

The NATO ASI directors worked alongside, and offer deep gratitude to Prof. Sir Tom Blundell, Director of the International School of Crystallography, Dr Colin Groom, Dr Neera Borkakoti, Dr John Irwin and Prof. Lodovico Riva di Sanseverino, who were in turn supported by a number of local facilitators.

The course was financed by NATO as an Advanced Study Institute. Additional support was given by the European Crystallographic Association, the International Union of Biochemistry and Molecular Biology, the International Union of Crystallography, the University of Bologna, AstraZeneca, Roche, Merck & Co., Boehringer Ingelheim, Bruker Corporation, Douglas Instruments, Informa UK, the Department of Pharmaceutical Chemistry, TTP Lab Tech, University of California at San Francisco.

Joel L. Sussman and Paola Spadon

LIST OF CONTRIBUTORS

Chris Abell
University Chemical Laboratory, Lensfield Road, Cambridge CB2 1EW, UK

Yasmin Aristei
Laboratory for Chemometrics and Cheminformatics, Chemistry Department, University of Perugia, Via Elce di sotto 10, Perugia, Italy

Edward N. Baker
Maurice Wilkins Center for Molecular Biodiscovery and School of Biological Sciences, University of Auckland, Auckland, New Zealand

Terese Bergfors
Department of Cell and Molecular Biology, Uppsala University, Biomedical Center Box 596, 751 24 Uppsala, Sweden

Tom L. Blundell
Department of Biochemistry, University of Cambridge, Tennis Court Road, Cambridge CB2 1GA, UK

Karen Bromfield
University Chemical Laboratory, Lensfield Road, Cambridge CB2 1EW, UK

Manfred Burghammer
European Synchrotron Radiation Facility, Grenoble, France

Emanuele Carosati
Laboratory for Chemometrics and Cheminformatics, Chemistry Department, University of Perugia, Via Elce di sotto 10, Perugia, Italy

Mario Centeleghe
Novartis Institutes for Biomedical Research, CH-4056 Basel, Switzerland

Vadim Cherezov
Department of Molecular Biology, The Scripps Research Institute, La Jolla, CA 92037, USA

Hee-Jung Choi
Departments of Molecular & Cellular Physiology and Structural Biology, Stanford University School of Medicine, 279 Campus Drive West, Stanford, CA 94305, USA

Alessio Ciulli
University Chemical Laboratory, Lensfield Road, Cambridge CB2 1EW, UK

Sandra W. Cowan-Jacob
Novartis Institutes for Biomedical Research, CH-4056 Basel, Switzerland

Gabriele Cruciani
Laboratory for Chemometrics and Cheminformatics, Chemistry Department, University of Perugia, Via Elce di sotto 10, Perugia, Italy

Peter W. Day
Departments of Molecular & Cellular Physiology, Stanford University School of Medicine, 279 Campus Drive West, Stanford, CA 94305, USA

Marcio V.B. Dias
Department of Biochemistry, University of Cambridge, Tennis Court Road, Cambridge CB2 1GA, UK

Patricia c. Edwards
MRC Laboratory of Molecular Biology, Cambridge, UK

Narayanan Eswar
Departments of Biopharmaceutical Sciences and Pharmaceutical Chemistry, California Institute for Quantitative Biosciences, University of California at San Francisco, San Francisco, CA, USA

Doriano Fabbro
Novartis Institutes for Biomedical Research, CH-4056 Basel, Switzerland

Gabriele Fendrich
Novartis Institutes for Biomedical Research, CH-4056 Basel, Switzerland

Robert f. Fischetti
Biosciences Division, Argonne National Laboratory, IL, USA

Andreas Floersheimer
Novartis Institutes for Biomedical Research, CH-4056 Basel, Switzerland

Juan J. Fung
Departments of Molecular & Cellular Physiology, Stanford University School of Medicine, 279 Campus Drive West, Stanford, CA 94305, USA

Pascal Furet
Novartis Institutes for Biomedical Research, CH-4056 Basel, Switzerland

Laura Goracci
Laboratory for Chemometrics and Cheminformatics, Chemistry Department, University of Perugia, Via Elce di sotto 10, Perugia, Italy

Michael A. Hanson
Department of Molecular Biology, The Scripps Research Institute, La Jolla, CA 92037, USA

Timo J. Heikkila
Department of Biochemistry, University of Cambridge, Tennis Court Road, Cambridge CB2 1GA, UK

Eran Hodis
Department of Structural Biology and The Israel Structural Proteomics Center, Weizmann Institute of Science, Rehovot, Israel

Nigel Howard
University Chemical Laboratory, Lensfield Road, Cambridge CB2 1EW, UK

Haejoo Kang
Maurice Wilkins Center for Molecular Biodiscovery and School of Biological Sciences, University of Auckland, Auckland, New Zealand

Gerhard Klebe
Institute of Pharmaceutical Chemistry, University of Marburg, Marbacher Weg 6, D35032 Marburg, Germany

Brian K. Kobilka
Departments of Molecular & Cellular Physiology, Stanford University School of Medicine, 279 Campus Drive West, Stanford, CA 94305, USA

Tong Sun Kobilka
Departments of Molecular & Cellular Physiology, Stanford University School of Medicine, 279 Campus Drive West, Stanford, CA 94305, USA

Peter Kuhn
Department of Molecular Biology and Department of Cell Biology, The Scripps Research Institute, La Jolla, CA 92037, USA

Janis Liebetanz
Novartis Institutes for Biomedical Research, CH-4056 Basel, Switzerland

Paul W. Manley
Novartis Institutes for Biomedical Research, CH-4056 Basel, Switzerland

Joseph Marcotrigiano
Department of Chemistry and Chemical Biology, Center for Advanced Biotechnology and Medicine, Rutgers University, NJ, USA

Asna Masood
Medarex, Inc., 521 Cottonwood Drive, Milpitas, CA 95035, USA

David Osborne
University Chemical Laboratory, Lensfield Road, Cambridge CB2 1EW, UK

Charles Parnot
Departments of Molecular & Cellular Physiology, Stanford University School of Medicine, 279 Campus Drive West, Stanford, CA 94305, USA

Jaime Prilusky
The Israel Structural Proteomics Center and Biological Services Unit, Weizmann Institute of Science, Rehovot, Israel

Thomas Proft
Maurice Wilkins Center for Molecular Biodiscovery and School of Medical Sciences, University of Auckland, Auckland, New Zealand

Søren G.F. Rasmussen
Departments of Molecular & Cellular Physiology, Stanford University School of Medicine, 279 Campus Drive West, Stanford, CA 94305, USA

Venkata R.P. Ratnala
Departments of Molecular & Cellular Physiology, Stanford University School of Medicine, 279 Campus Drive West, Stanford, CA 94305, USA

Paul Rheinberger
Novartis Institutes for Biomedical Research, CH-4056 Basel, Switzerland

Daniel K. Rohrer
Medarex, Inc., 521 Cottonwood Drive, Milpitas, CA 95035, USA

Daniel M. Rosenbaum
Departments of Molecular & Cellular Physiology, Stanford University School of Medicine, 279 Campus Drive West, Stanford, CA 94305, USA

Gabriele Rummel
Novartis Institutes for Biomedical Research, CH-4056 Basel, Switzerland

Andrej Sali
Departments of Biopharmaceutical Sciences and Pharmaceutical Chemistry, California Institute for Quantitative Biosciences, University of California at San Francisco, San Francisco, CA, USA

Ruslan Sanishvili
Biosciences Division, Argonne National Laboratory, IL, USA

Giovanna Scapin
Departments of Global Structural Biology, Merck & Co., Inc., PO BOX 2000, Rahway NJ 07065,USA. Current address: Schering-Plough Research Institute, 2015 Galloping Hill Road K15-1-1800, Kenilworth, NJ 07033, USA

Gebhard F.X. Schertler
MRC Laboratory of Molecular Biology, Cambridge, UK

Duncan Scott
University Chemical Laboratory, Lensfield Road, Cambridge CB2 1EW, UK

Israel Silman
The Israel Structural Proteomics Center and Neurobiology Department, Weizmann Institute of Science, Rehovot, Israel

Hernani L. Silvestre
Department of Biochemistry, University of Cambridge, Tennis Court Road, Cambridge CB2 1GA, UK

Raymond C. Stevens
Department of Molecular Biology, The Scripps Research Institute, La Jolla, CA 92037, USA

Sachin Surade
Department of Biochemistry, University of Cambridge, Tennis Court Road, Cambridge CB2 1GA, UK

Joel L. Sussman
Department of Structural Biology and The Israel Structural Proteomics Center, Weizmann Institute of Science, Rehovot, Israel

Foon Sun Thian
Departments of Molecular & Cellular Physiology, Stanford University School of Medicine, 279 Campus Drive West, Stanford, CA 94305, USA

Janet M. Thornton
EMBL-European Bioinformatics Institute, Wellcome Trust Genome Campus, Hinxton, Cambridge CB10 1SD, UK

James D. Watson
EMBL-European Bioinformatics Institute, Wellcome Trust Genome Campus, Hinxton, Cambridge CB10 1SD, UK

Alvin Hung Wei
University Chemical Laboratory, Lensfield Road, Cambridge CB2 1EW, UK

William I. Weis
Departments of Molecular & Cellular Physiology and Structural Biology, Stanford University School of Medicine, 279 Campus Drive West, Stanford, CA 94305, USA

Shijun Wen
University Chemical Laboratory, Lensfield Road, Cambridge CB2 1EW, UK

Xiao-Jie Yao
Departments of Molecular & Cellular Physiology, Stanford University School of Medicine, 279 Campus Drive West, Stanford, CA 94305, USA

Ada Yonath
Department of Structural Biology, Weizmann Institute, Rehovot, Israel

SURFACE PROTEINS OF GRAM-POSITIVE PATHOGENS: USING CRYSTALLOGRAPHY TO UNCOVER NOVEL FEATURES IN DRUG AND VACCINE CANDIDATES

EDWARD N. BAKER[1*], THOMAS PROFT[2], HAEJOO KANG[1]
Maurice Wilkins Center for Molecular Biodiscovery and [1]School of Biological Sciences, University of Auckland, Auckland, New Zealand; [2]School of Medical Sciences, University of Auckland, Auckland, New Zealand

Abstract. Proteins displayed on the cell surfaces of pathogenic organisms are the front-line troops of bacterial attack, playing critical roles in colonization, infection and virulence. Although such proteins can often be recognized from genome sequence data, through characteristic sequence motifs, their functions are often unknown. One such group of surface proteins is attached to the cell surface of Gram-positive pathogens through the action of sortase enzymes. Some of these proteins are now known to form pili: long filamentous structures that mediate attachment to human cells. Crystallographic analyses of these and other cell surface proteins have uncovered novel features in their structure, assembly and stability, including the presence of inter- and intramolecular isopeptide crosslinks. This improved understanding of structures on the bacterial cell surface offers opportunities for the development of some new drug targets and for novel approaches to vaccine design.

Keywords: Gram-positive pathogens; Group A streptococcus; cell surface proteins; sortases; bacterial pili; pilus assembly; genome analysis; electron microscopy; X-ray crystallography; mass spectrometry; isopeptide bonds; phylogenetic analysis; vaccine development

* To whom correspondence should be addressed. Edward N. Baker, Maurice Wilkins Center for Molecular Biodiscovery, School of Biological Sciences, University of Auckland, Auckland, New Zealand; e-mail: ted.baker@auckland.ac.nz

J.L. Sussman and P. Spadon (eds.), *From Molecules to Medicines*,

1. Introduction

Proteins that are secreted from pathogenic organisms or are displayed on their cell surfaces provide a front line of attack for disabling host defenses or mediating colonization and infection of host tissues. Their prominence as antigens and ready accessibility also makes them prime candidates for drug or vaccine design. Most of these proteins remain uncharacterized, but the increasing availability of complete genome sequences for important human and animal pathogens provides exciting new opportunities for the discovery of new therapeutic possibilities.

Secreted or cell-surface proteins can often be recognized from the presence of characteristic sequence motifs, such as the N-terminal signal sequences that identify potential secreted or lipid-anchored proteins, and the C-terminal sequences for glycosylphosphatidylinositol (GPI) anchors.[1] In Gram-positive bacteria, cysteine transpeptidase enzymes called sortases mediate attachment of substrate proteins to the cell wall, for display on the cell surface.[2] The substrate proteins can be recognized by a sortase recognition sequence, typically LPxTG (x = any amino acid), immediately preceding a C-terminal hydrophobic region. Sortase action cleaves the substrate protein after the Thr residue of this motif, and joins the new terminal carboxyl group to an amino group of the cell wall peptidoglycan via a covalent isopeptide bond. In this manner, general sortases mediate the attachment of many cell surface proteins, while some specialized sortases attach certain specific substrate proteins.

Group A streptococci (GAS), the cause of human throat and skin infections, as well as serious invasive diseases such as necrotizing fasciitis and toxic shock[3] has three sortases and 17 predicted substrate proteins,[4] expected to be displayed on the cell surface. Some of these are encoded in a pathogenicity island called the FCT (fibronectin-binding, collagen-binding, T-antigen) region[5] (Fig. 1). This varies from one strain to another, but in every case encodes at least one sortase (sometimes two) and a number of substrate proteins. The latter were mostly of unknown function, until the discovery that a gene cluster within the FCT region encodes proteins involved in the formation of pilus structures on the cell surface.[6]

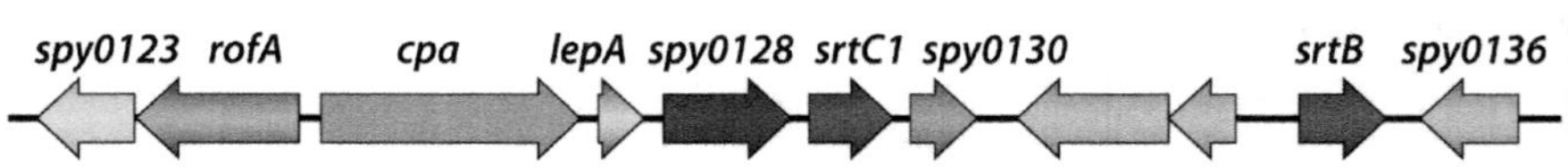

Figure 1. The FCT region from the M1 strain of GAS. Both *srtC1* and *srtB* encode sortases, and *cpa*, *spy0128* and *spy0130* encode pilus subunits.

2. Bacterial pili

Bacterial pili are long, thin filamentous structures that extend from the bacterial cell surface and mediate host cell adhesion, biofilm formation and other aspects of colonization. The pili found on Gram-negative organisms are well characterized, particularly the Type IV pili from *E. coli* and Pseudomonas and Neisseria species and the Type P and Type 1 pili of uropathogenic *E. coli* strains. These are long (1–4 μm), thin (5–8 nm) and flexible, but are also extremely strong, able to withstand extreme physical stresses. Gram-negative pili form by the non-covalent association of pilin subunits. Type IV pili are based on superhelical assemblies of subunits held together by hydrophobic interactions.[7] Type 1 and P pili, in contrast, assemble by a process of donor strand exchange in which incomplete immunoglobulin (Ig)-type subunits are completed and linked by the insertion of a strand from the next subunit in the assembly.[8,9]

Surprisingly, the pili on Gram-positive organisms have gone largely unrecognized until very recently, probably because they are extremely thin (2–3 nm) and often not visible by conventional negative staining EM techniques. The availability of genome sequence data, enabling candidate pilus proteins to be identified, has changed this, permitting immunogold labeling and EM visualization of pili on important pathogenic species such as *Corynebacterium diphtheriae*,[10] *Streptococcus pneumoniae*[11] and Group A and B streptococci.[6,12,13] The assembly of these Gram-positive pili depends on sortase enzymes, which elongate the pilus oligomer by progressive addition of backbone subunits; each subunit is cleaved at its C-terminal sortase recognition motif, and the new C-terminus joined to a lysine ε-amino group on the next subunit *via* an isopeptide bond. The entire assembly is then joined to the cell wall peptidoglycan, again by sortase action.[14,15] For both *C. diphtheriae* and *Bacillus cereus* it has been shown that two different sortases are required, a specific sortase which catalyzes formation of the pilus oligomer and a general (house-keeping) sortase for the final cell wall attachment.[16,17] Gram-positive pili thus differ fundamentally from Gram-negative pili in both their mode of assembly and the covalent linkages between subunits.

This mode of enzyme-mediated conjugation has its counterparts in other biological processes such as ubiquitination[18] and transglutamination,[19] in which the ε-amino groups of lysine residues on particular proteins are joined *via* isopeptide bonds to either a C-terminal carboxyl group or a glutamine side chain, respectively, on another protein, generating conjugated proteins.

3. Role of crystallography

X-ray crystallography plays a unique role in the discovery of structure and function. It can stimulate entirely new functional hypotheses. It can provide the high resolution necessary to define biological mechanisms or binding specificity, critical for drug development. The discovery of bound small molecules or ions in a crystal structure can provide unexpected functional insights, and the ways molecules pack in a crystal can identify surfaces that may be used in transient protein-protein interactions in the cell. It may be relevant here that the protein concentration in a crystal (about 500 mg/mL) is not too different from the concentration of macromolecules in the crowded intracellular environment, estimated at 300–400 mg/mL.[20]

Importantly, crystallography also has the power to discover the completely unexpected and to do so with a high degree of certainty. Examples from chemistry include the definition of the β-lactam ring of penicillin and the first ever discovery of a metal-carbon bond – identified in the structure of the B_{12} coenzyme.[21] A similar example from biology is the discovery of an unprecedented covalent bond between Cys and Tyr side chains in the structure of galactose oxidase.[22] Two aspects of our investigations of bacterial pili emphasize these points.

4. Structure and assembly of GAS pili

Pilus structure and assembly in GAS (*S. pyogenes*) depends on the products of four genes: a sortase SrtC1; a major pilin subunit which forms the polymeric pilus backbone; and two associated minor pilins which decorate the pilus structure. In the M1 strain of GAS these are Spy0129, Spy0128, Spy0125 and Spy0130, respectively.[6]

The crystal structure of the sortase SrtC1, solved at 2.3 Å resolution (H. Kang, unpublished), showed that it has the archetypal sortase fold seen in *Staphylococcus aureus* SrtB,[23] and shares the same catalytic residues. On the other hand, it has large insertions in portions of the structure that help form the substrate-binding region. This is consistent with the idea that SrtC1 is a specialized sortase required specifically for pilus assembly, and correlates with the different sortase motif in the backbone pilin subunit: EVPTG, compared with the generic LPxTG.

Determination of the structure of the major pilin protein Spy0128, by X-ray crystallography at 2.2 Å resolution,[24] shows that it is folded into two immunoglobulin-like (Ig-like) domains, to give an elongated molecule approximately 100 Å long and 20–30 Å wide (Fig. 2). The fold of each domain resembles the "inverse IgG" fold found in the repeating CnaB domains of the collagen-binding adhesin Cna from *S. aureus*[25] and demonstrates a possible evolutionary relationship that links pili to a large family of cell surface proteins involved in binding to the extracellular matrix.

In addition to the monomer structure, the crystal structure brought two unexpected surprises, however, that give important insights into pilus structure and assembly. Firstly, the crystal asymmetric unit contains three independent molecules that generate columns of molecules extending through the crystal. This arrangement, which is also seen in two different crystal forms, provides a very persuasive model for pilus assembly. Successive molecules pack head-to-tail, with the C-terminus of one molecule close to an invariant lysine residue (Lys161) in the N-terminal domain of the preceding molecule in the column. Mass spectrometry of native pili extracted from GAS[24] showed that Lys161 does indeed form an isopeptide bond with the C-terminus of the next molecule following sortase cleavage, validating this model.

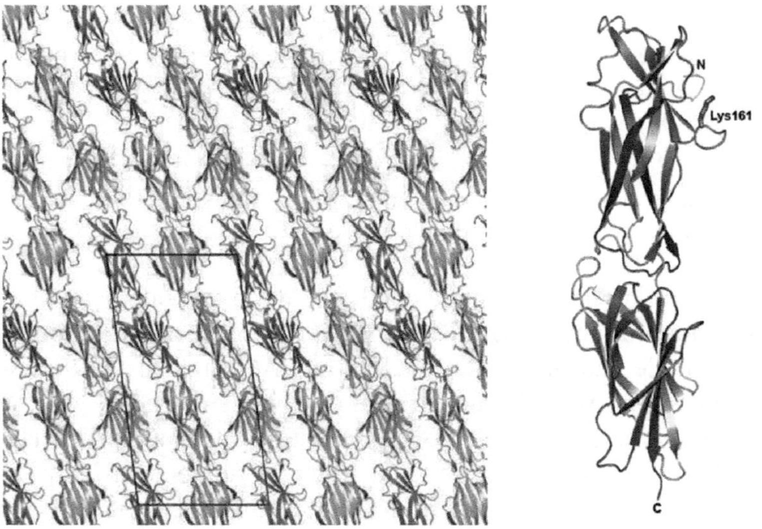

Figure 2. Crystal structure of GAS major pilin Spy0128. (a) Crystal packing, showing columns of molecules that model pilus assembly. (b) Ribbon diagram of the Spy0128 monomer. The position of Lys161, is joined to the C-terminus of the next molecule by sortase action, is shown.

A second surprise came with the discovery of two internal cross-links, one in each domain of the Spy0128 monomer. In both cases a lysine side chain is joined to an asparagine side chain *via* a covalent isopeptide bond, with an adjacent glutamic acid residue playing an essential part in the reaction; mutation of this glutamic acid residue abolishes isopeptide bond formation. The isopeptide bonds were clearly indicated by continuous electron density, extending from the lysine ε-amino group into the asparagine carboxyamide group, and were subsequently verified by mass spectrometry.[24]

The bonds clearly form during protein folding, as the Lys, Asn and Glu residues become sequestered in a hydrophobic environment. A similar example of self-generated isopeptide bonds – albeit intermolecular – had been documented just once before, in the "chain-mail" capsid structure of the bacteriophage HK97.[26]

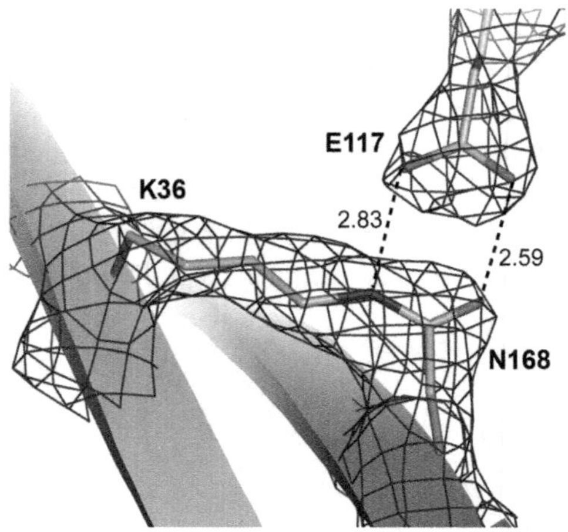

Figure 3. Isopeptide bond in the N-terminal domain of Spy0128, showing the continuous electron density for the side chains of Lys36 and Asn168.

The possibility that isopeptide bonds could provide internal crosslinks in a protein, joining amino acid side chains, has been raised before, but no proven example had previously been found. A search of the Protein Data Bank using a Lys-Asn-Glu/Asp structural template does, however, identify several cases where similar Lys-Asn isopeptide bonds almost certainly are present, but have been missed in the structure refinement.[24] The proteins involved are all surface proteins from Gram-positive bacteria, one of them a minor pilin,[27] suggesting that such structures may be more common.

It seems likely that the internal isopeptide bonds seen in GAS pili may be a common feature of Gram-positive pili. The backbone pilin subunits have been identified for a number of other pathogens: Group B streptococcus,[13] *C. diphtheriae*,[10] *S. pneumoniae*[11] and *B. cereus*.[17] These are, however, extremely variable in size and sequence, making realistic sequence alignments with GAS impossible. Preliminary attempts to use secondary structure prediction to guide sequence alignments do suggest that the major pilins from all these species possess a common framework, albeit with additional domains in some cases. For *B. cereus*, for example, the major pilin comprises three Ig-like domains and mass spectrometry has shown that these also contain internal isopeptide bonds.[28]

5. Role of isopeptide bonds in GAS pili

The width of GAS pili, as seen by electron microscopy, is only of the order of 20–30 Å,[6] equivalent to the width of a single molecule. While it would be possible for several chains of molecules to wind around each other in a coiled coil, as is suggested for the pili of *S. pneumoniae*,[29] this does not seem likely for GAS, given the polar surface of the subunits. The intermolecular isopeptide bonds that generate the covalent polymer must therefore be essential for giving strength and stability to these long, thin assemblies.

The internal isopeptide bonds are especially interesting since these are the first such internal crosslinks to be discovered. They are strategically placed, in each case linking the first and last β-strands of the domain, such as to provide maximum resistance to unfolding from tensile forces applied along the pilus direction.[30] Unfolding studies on mutant Spy0128 proteins show that the loss of one or both isopeptide bonds severely reduces thermal stability (H. Kang, unpublished), and proteolysis experiments show that the isopeptide bonds confer considerable protection against digestion by proteases.[24] In this context, we note that the backbone pilins of GAS correspond to the classic Lancefield T-antigens, originally named for their trypsin (T) resistance.

6. Implications for drug and vaccine design

Cell-surface proteins present no barriers for drug access, and are easily "seen" by antibodies, making them attractive targets for drug or vaccine development. GAS pili, for example, have been shown by gene knockout to be required for adhesion to human skin and tonsil cells, and for bacterial colonisation.[31] The sortase SrtC1 which is essential for pilus assembly is one obvious drug target, as is the adhesin through which the pili attach to human cells – although this has yet to be identified unequivocally or defined structurally. Another potential target is a highly conserved surface on the GAS major pilin protein, where the sortase reaction occurs, as this could block pilus assembly. Small molecules targeted against Type 1 pili of uropathogenic *E. coli* have given proof of concept for this approach.[32] An even more powerful approach would be a generic inhibitor for sortase enzymes, since this would potentially affect all the sortase-anchored surface proteins on the bacterium.

Various studies have shown that immunisation with pilus components can give protection against bacteria expressing those pili.[6,13,14,33] To date, this approach has not yet resulted in effective vaccines, because of sequence variations in the pili between different strains. Effective vaccines may, however, be generated by mapping these variations on to the structures of the pilin proteins, and using combinations of the predominant pilus types. For diseases such as tuberculosis, where an improved vaccine is desperately needed, the recent discovery that *Mycobacterium tuberculosis*, too, produces pili during human infection[34] points to a new possible way forward.

References

1. Eisenhaber, F., Eisenhaber, B., Kubina, W., Maurer-Stroh, S., Neuberger, G., Schneider, G. & Wildpaner, M. (2003). *Nucl. Acids Res.* **31**, 3631–3634.
2. Mazmanian, S.K., Liu, G., Ton-That, H. & Schneewind, O. (1999). *Science* **285**, 760–763.
3. Cunningham, M.W. (2000). *Clin. Microbiol. Rev.* **13**, 470–511.
4. Rodriguez-Ortega, M.J., Norais, N., Bensi, G., Liberatori, S., Capo, S., Mora, M. et al. (2006). *Nature Biotechnol.* **2**, 191–197.
5. Kreikemeyer, B., Klenk, M. & Podbielski, A. (2004). *Int. J. Med. Microbiol.* **294**, 177–188.
6. Mora, M., Bensi, G., Capo, S., Falugi, F., Zingaretti, C., Manetti, A.G.O. et al. (2005). *Proc. Natl. Acad. Sci. USA* **102**, 15641–15646.
7. Craig, L., Pique, M.E. & Tainer, J.A. (2004). *Nature Rev. Microbiol.* **2**, 363–378.
8. Sauer, F.G., Futterer, K., Pinkner, J.S., Dodson, K.W., Hultgren, S.J. & Waksman, G. (1999). *Science* **285**, 1058–1061.
9. Vetsch, M., Puorger, C., Spirig, T., Grauschopf, U., Weber-Ban, E.U. & Glockshuber, R. (2004). *Nature* **431**, 329–332.
10. Ton-That, H. & Schneewind, O. (2003). *Mol. Microbiol.* **50**, 1429–1438.
11. Barocchi, M.A., Ries, J., Zogaj, X., Hemsley, C., Albiger, B., Kanth, A. et al. (2006). *Proc. Natl. Acad. Sci. USA* **103**, 2857–2862.
12. Lauer, P., Rinaudo, C.D., Soriani, M., Margarit, I., Maione, D., Rosini, R. et al. (2005). *Science* **309**, 105.
13. Rosini, R., Rinaudo, C.D., Soriani, M., Lauer, P., Mora, M., Maione, D. et al. (2006). *Mol. Microbiol.* **61**, 126–141.
14. Telford, J.L., Barocchi, M.A., Margarit, I., Rappuoli, R. & Grandi, G. (2006). *Nature Rev. Microbiol.* **4**, 509–519.
15. Scott, J.R. & Zahner, D. (2006). *Mol. Microbiol.* **62**, 320–330.
16. Swaminathan, A., Mandlik, A., Swierczinski, A., Gaspar, A., Das, A. & Ton-That, H. (2007). *Mol. Microbiol.* 66, 961–974.
17. Budzik, J.M., Marraffini, L.A. & Schneewind, O. (2007). *Mol. Microbiol.* **66**, 495–510.
18. Pickart, C.M. (2001). *Annu. Rev. Biochem.* 70, 503–533.
19. Greenberg, C.S., Birckbichler, P.J. & Rice, R.H. (1991). *FASEB J.* 5, 3071–3077.
20. Zimmerman, S.B. & Trach, S.O. (1991). *J. Mol. Biol.* **222**, 599–620.
21. Hodgkin, D.C. (1965). *Les Prix Nobel*, 157–178.
22. Ito, N., Phillips, S.E.V., Stevens, C., Ogel, Z.B., McPherson, M.J., Keen, J.N. et al. (1991). *Nature* **350**, 87–90.
23. Zhang, R., Wu, R., Joachimiak, G., Mazmanian, S.K., Missiakis, D.M., Gornicki, P. et al. (2004). *Structure* **12**, 1147–1156.
24. Kang, H.J., Coulibaly, F., Clow, F., Proft, T. & Baker, E.N. (2007). *Science* **318**, 1625–1628.
25. Deivanayagam, C.C.S., Rich, R.L., Carson, M., Owens, R.T., Danthuluri, S., Bice, T. et al. (2000). *Structure* **8**, 67–78.

26. Wikoff, W.R., Liljas, L., Duda, R.L., Tsuruta, H., Hendrix, R.W. & Johnson, J.E. (2000). *Science* **289**, 2129–2133.
27. Krishnan, V., Gaspar, A.H., Ye, N., Mandlik, A., Ton-That, H. & Narayana, S.V.L. (2007). *Structure* **15**, 893–903.
28. Budzik, J.M., Marraffini, L.A., Souda, P., Whitelegge, J.P., Faull, K.F. & Schneewind, O. (2008). *Proc. Natl. Acad. Sci. USA* **105**, 10215–10220.
29. Hilleringmann, M., Giusti, F., Baudner, B.C., Masignani, V., Covacci, A., Rappuoli, R. et al. (2008). *PLOS Pathogens* **4**, 1–11.
30. Yeates, T.O. & Clubb, R.T. (2007). *Science* **318**, 1558–1559.
31. Abbot, E.L., Smith, W.D., Siou, G.P.S., Chiriboga, C., Smith, R.J., Wilson, J.A. et al. (2007). *Cell. Microbiol.* **9**, 1822–1833.
32. Pinkner, J.S., Remaut, H., Buelens, F., Miller, E., Aberg, V., Pemberton, N. et al. (2006). *Proc. Natl. Acad. Sci. USA* **103**, 17897–17902.
33. Buccato, S., Maione, D., Rinaudo, C.D., Volpini, G., Taddei, A.R., Rosini, R. et al. (2006). *J. Infect. Dis.* **194**, 331–340.
34. Alteri, C.J., Xicohtencatl-Cortes, J., Hess, S., Caballero-Olin, G., Giron, J.A. & Friedman, R.L. (2007). *Proc. Natl. Acad. Sci. USA* **104**, 5145–5150.

THE RAPID CRYSTALLIZATION STRATEGY FOR STRUCTURE-BASED INHIBITOR DESIGN

TERESE BERGFORS*
Department of Cell and Molecular Biology, Uppsala University, Biomedical Center, Box 596, 751 24 Uppsala, Sweden

Abstract. RAPID (Rapid Approaches to Pathogen Inhibitor Discovery) is an integrated center for structural biology, computational chemistry, and medicinal chemistry at Uppsala University, Sweden. The main target of the structural biology section is *Mycobacterium tuberculosis*. Key concepts in the crystallization strategy include minimal screening and buffer optimization. Examples are presented showing how these concepts have been successful in RAPID projects. Three screening methods are used: vapor-diffusion, micro-batch, and microfluidics. Our experiences may be relevant for other small, academic laboratories involved in structure-based inhibitor design.

Keywords: Buffer effects, crystallization strategy, manual screening, *Mycobacterium tuberculosis*, protein crystallization, protein–inhibitor complexes, storage of protein, structure–based inhibitor design

1. Introduction

RAPID stands for Rapid Approaches to Pathogen Inhibitor Discovery; it is an integrated center for structural biology, computational chemistry, and medicinal chemistry at Uppsala University, Sweden. The goal of RAPID is structure-based inhibitor design against proteins from the micro-organisms that cause tuberculosis, malaria, leishmaniasis and trypanosomiasis. The structural biology section focuses on tuberculosis, which is caused by *Mycobacterium tuberculosis*.

---

* To whom correspondence should be addressed. Terese Bergfors, Department of Cell and Molecular Biology, Uppsala University, Biomedical Center Box 596, 751 24 Uppsala, Sweden; e-mail: Terese.Bergfors@icm.uu.se

The three sections of RAPID interact closely with each other and with their industrial partners. The structural biology section performs target selection, cloning of the gene, expression and purification of the protein, followed by crystallization screening, data collection and structure determination of the proteins and protein-inhibitor complexes. The medicinal/combinatorial chemistry section synthesizes and optimizes the inhibitors for the structural biology section. This chemistry section also performs enzyme inhibition assays and metabolic stability tests. The third section of RAPID is comprised of the computational chemists who perform homology-based modeling, virtual screening, library design, docking routines, scoring functions, and ADME (adsorption, distribution, metabolism and excretion) prediction.

RAPID has been funded since January 2003 by the Swedish Foundation for Strategic Research. The structural biology section has deposited 22 structures from *Mycobacterium tuberculosis* in the PDB (see Table 1); 10 of these are protein–inhibitor or protein–ligand complexes. The structural biology section employs ten graduate students and four principal investigators (PIs).

TABLE 1. Deposited *M. tuberculosis* structures from RAPID 2003–2007.

Rv	Protein	PDB ID
Rv0009	Peptidyl-prolyl cis-trans isomerase A	1W74
Rv0130	Conserved hypothetical	2C2I
Rv0216	Conserved hypothetical	2BI0
Rv1284	β-carbonic anhydrase related protein	1YLK
Rv1295	Threonine synthase	2D1F
Rv2220	Glutamine synthetase	2BVC
Rv2461c	ClpP1	2C8T
Rv2465c	Ribose-5-phosphate isomerase B	1USL
	Ribose-5-phosphate isomerase B	2BES
	Ribose-5-phosphate isomerase B	2BET
Rv2740	Epoxide hydrolase	2BNG
Rv2870c	1-deoxy-D-xylulose 5-phosphate reductoisomerase	2C82
	1-deoxy-D-xylulose 5-phosphate reductoisomerase	2JCV
	1-deoxy-D-xylulose 5-phosphate reductoisomerase	2JCX
	1-deoxy-D-xylulose 5-phosphate reductoisomerase	2JCY
	1-deoxy-D-xylulose 5-phosphate reductoisomerase	2JCZ
	1-deoxy-D-xylulose 5-phosphate reductoisomerase	2JD0
	1-deoxy-D-xylulose 5-phosphate reductoisomerase	2JD1
	1-deoxy-D-xylulose 5-phosphate reductoisomerase	2JD2
Rv3588c	β-carbonic anhydrase (dimer)	1YM3
	β-carbonic anhydrase (tetramer)	2A5V
Rv3778c	Possible oxido-reductase	3CAI

The PIs are responsible for their particular area (protein expression, crystallization, methods development, structure solution) whereas students are trained in the entire process, from cloning to structure refinement. The two chemistry sections together comprise 14 scientists and students. As the PI responsible for crystallization, I will focus below on the crystallization strategy within the structural biology section.

2. Materials and methods

2.1. PROTEIN PRODUCTION

After target selection, the gene is cloned into a pCR®T7/CT-TOPO® or pEXP5-CT/TOPO® vector (Invitrogen). Each construct carries an N-terminal 6-His tag without a linker. The His-tag is not removed for the crystallization trials. The plasmid is transformed into *Escherichia coli* TOP10 cells (Invitrogen). Positive clones are sequenced to confirm correctness, after which they are used to transform *E. coli* strain BL21/AI. Cultures are grown in 2.8 L Buchner flasks containing 1 L LB-medium supplemented with ampicillin, grown to log phase, then induced with 0.02% arabinose. Growth at 37°C continues a further 2–4 h before harvesting by centrifugation. Cell pellets not processed immediately are stored at –20°C.

The standard lysis buffer is 20 mM Tris-HCl, pH 8.0, 100 mM NaCl, 0.1% Triton X-100. After cell debris is removed by centrifugation, the supernatant is applied to a Ni-IMAC column (Qiagen) and the His-tagged protein is eluted with an imidazole gradient. The second purification step is size-exclusion chromatography (SEC), usually a Superdex 75 column, (GE Healthcare), equilibrated in 10 mM Tris-HCl, pH 8.0 and 150 mM NaCl. The protein is always assayed by SDS-PAGE and sometimes with native PAGE as well.

The pure protein fractions are pooled and concentrated at 15°C in centrifugal concentration devices (VivaSciences). The choice of buffer can be critical to the outcome of the concentration step. The protein is eluted in the SEC buffer, which serves as the default buffer in the concentration step. Centrifugation is paused every 5 min to monitor the behavior of the protein. Should the protein show signs of precipitation, the centrifugation step is discontinued, the protein solution (supernatant) is cleared of precipitate, and the supernatant is tested in a buffer screen. The buffer screen is performed as a vapor-diffusion setup where the experimental droplet consists of a 1:1 mixture of protein and reservoir solutions. The reservoir solutions in this case do not contain precipitants but only buffers, from pH 3.5 to 10.5 at concentrations of 100 mM. The droplet is equilibrated over the reservoir for 1 day or longer and observed for signs of precipitation. To be able to see the precipitation, the protein concentration needs to be high enough. Therefore I recommend using a concentration from 3 to 10 mg/mL, but it can be lower if the protein is already precipitating. Here the goal of the experiment is to find buffer conditions where the droplet remains clear, i.e., the protein remains soluble. The SEC buffer is then exchanged, by diafiltration or dialysis, for one of those found in the screen. The concentration by centrifugation step can be resumed after this buffer exchange – with the aim of achieving a concentration from 3 to 25 mg/mL for the crystallization screen. This method, under the name Optimum Solubility Screening, as well as variations of it, have been recently described in the literature.[1–3] There are now commercial buffer screens available for this purpose (Jena BioSciences, Molecular Dimensions, etc.).

After the protein is concentrated, it is immediately submitted to crystallization screening. Any surplus protein is flash-frozen according to the protocol developed in the laboratory of Prof. Wim Hol.[4]

2.2. CRYSTALLIZATION SCREENING

Crystallization screening is performed on an Oryx 6 robot (Douglas Instruments, UK) as sitting-drop vapor-diffusion trials with 100 μL precipitant solution in the reservoirs and drop volumes of 150 nL protein and 150 nL precipitant. Two different screens kits are used: JCSG+ (available from Qiagen, Molecular Dimensions, etc.), containing 96 conditions, and Mini (Molecular Dimensions), with 24 conditions.

In parallel with the vapor-diffusion trials, the same screen is set up in two additional geometries: as microbatch experiments and in microfluidic chips, (Microlytic, Denmark, www.microlytic.com). The microfluidic setup is shown in Fig. 1. In the microbatch trials, the volumes are identical to those in the vapor-diffusion droplet. A 1:1 mixture of parafin:silicone oil is used to cover the microbatch droplets.

The setups are incubated at 20°C. Other temperatures (4°C, 27°C) might also be tested, but not until the second tier of experiments. The crystallization experiments are observed and the results are recorded in Xtrack, a laboratory information management system developed in our laboratory.[5] The setups are monitored immediately upon setup, then daily for a week, and thereafter on a weekly basis for about 3 months. Visual assessment and recording of the results are performed manually.

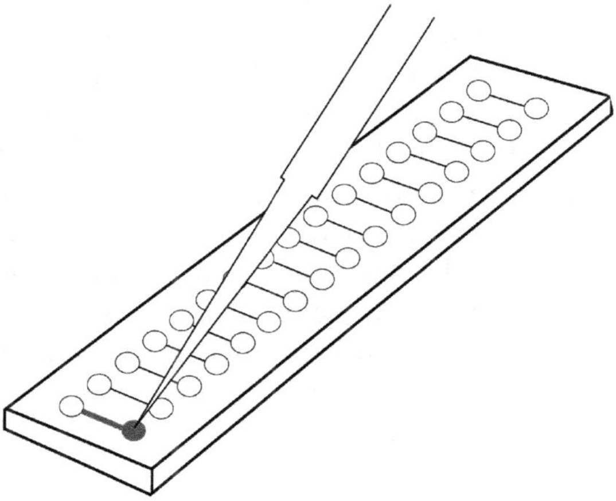

Figure 1. Sketch of the Crystal Former™ from Microlytic. Here a single channel pipette is used to fill the inlets; a multichannel pipette can be used for simultaneous filling of the inlets. The chip is SBS-compatible for robot loading. The 16 protein inlets are loaded with 150–400 nL each; the channels fill by capillarity. The precipitant is then added to the inlet at the opposite end of the channel. Both rows of inlets are sealed with tape or foil. The figure is reprinted with permission from Microlytic.

2.3. SECOND- AND THIRD-TIER SCREENING

There is no shortage of commercially available screening kits to try, should the first two fail to produce any promising leads. The second tier of experiments varies the temperature and protein concentration and may be expanded to include three other screens: Pact (Qiagen, Molecular Dimensions, etc.), Quik (a phosphate/pH screen, Hampton Research) and Silver Bullet (Hampton Research). The His-tag is still retained at this level of the screening.

Microseeding with any promising solid phase produced in the first round of screening is always done in the second-tier experiments. Promising solid phases include microcrystals, but even crystalline precipitates, spherulites, or seemingly amorphous precipitate. Many amorphous precipitates harbor some crystallinity which is not obvious in visual inspection through the microscope. A seed slurry is generated from the precipitate or other solid phase and a small fraction of it is included as an additive to the new drops. The procedure whereby seeds originating in one mother liquor are used to "innoculate" drops with unrelated mother liquors can be done robotically.[6] It has been dubbed "matrix seeding".[7]

If a third tier of experiments should be necessary, a new construct is made, sometimes without the His-tag. Our construct in the first and second tiers does not have a cleavage site for the His-tag. As a result, removal of the tag requires a second cloning step. However, all the structures in Table 1 were solved with N-terminal 6-His tags and we have not yet encountered any examples in the RAPID project where removal of the His tag was critical to obtaining the crystals. The most comprehensive analysis to date of His tags in the PDB concludes that they are generally benign.[8] For our project other deletions, usually from the N- and C-termini, have proved to be more effective than His-tag removal for making the proteins more "crystallizable". Certainly by this stage, if not already in the initial cloning step, the amino acid sequence of the protein is analyzed with the bioinformatics programs available at www.disprot.org for evidence of disordered regions that could interfere with crystallization. These are removed in the new constructs.

2.4. SCREENING OF PROTEIN–INHIBITOR COMPLEXES

Two methods for introducing an inhibitor into the protein are co-crystallization and soaking. Others are discussed in an excellent review.[9] In co-crystallization, the protein is incubated together with the inhibitor for a defined time and then the protein–inhibitor complex is set up in crystallization droplets. In the other method, soaking, the protein is crystallized without the inhibitor, and then the inhibitor is soaked into the protein crystal. There are advantages and disadvantages to both methods, but soaking is usually the easier of the two. However, the inhibitor may cause such a conformational change in the protein that the crystal contacts are disrupted. At RAPID, co-crystallization experiments on the protein–inhibitor complex are screened as described above. If crystals of the apo protein are available, they are used in soaking experiments with the ligands and for microseeding the co-crystallization experiments. Soaking experiments are performed in parallel with the co-crystallization ones (when apo crystals are available) to increase the chances of obtaining a crystal of the complex.

The limited solubility in aqueous buffers of the majority of the inhibitors in this project is a major complication, regardless of whether the binding attempts are made as co-crystallization or soaking experiments. The inhibitors are usually dissolved in neat (100%) DMSO (dimethyl sulfoxide). When the inhibitor is added to the protein solution (co-crystallization) or the mother liquor containing the crystal (soaking), the resulting dilution of the solvent (in this case, DMSO) leads to precipitation of the inhibitor. Enough inhibitor might still remain in solution to bind to the protein, but this is not known until the crystal structure is solved. Apart from solubility issues, the inhibitor binding is also dependent in varying degrees upon the buffer, pH, and other mother liquor components. Therefore, second-tier experiments may include exchanging the mother liquor before soaking or co-crystallization with the inhibitor.

3. Discussion

Three concepts in our crystallization strategy at RAPID are discussed more in depth below. These deal with the questions of how many conditions are "enough" to test in the crystallization screening; the protocol for storing and freezing the protein to improve reproducibility in the crystallization trials; and the role of the protein buffer.

3.1. THE CONCEPT OF MINIMAL SCREENING

Our initial screening strategy uses only 120 precipitant/mother liquor combinations (the 96 in the JCSG+ screening kit and the 24 in the Mini). These are applied to the protein in up to three different geometries: vapor-diffusion, microbatch, and microfluidics. The three geometries affect the equilibration kinetics so differently that each format can generate "hits" that are unique to it. We are currently compiling the success rates of the three geometries for our proteins, but so far our results show that each of the three geometries produce overlaps with each other as well as hits that are geometry-specific. Thus, with only two screening kits and three geometries, 360 conditions can be tested per protein concentration and temperature. The number of screening kits commercially available nowadays is enormous and maintaining an entire stock of them, reformatting them to Deep Well blocks, etc., are expensive and laborious tasks. For simplicity and cost-effectiveness, we therefore use only two screens in the first tier of experiments. At this stage the goal is not to obtain well-diffracting crystals, although that is a welcome side-effect, should it happen. Instead, the goal of the initial screen is to answer the question: "Is this protein likely to crystallize or not?" Extensive screening with hundreds and hundreds of conditions has a limited return on the investment it requires. The efficiency study by Segelke[10] showed that a screen consisting of 300 conditions is a reasonable enough size to determine if "a protein is likely to crystallize or not". Another study by Newman et al.[11] found similar results. The advantage of minimal screening in a first tier of experiments is that it may produce results with little investment of time and effort, but it does not preclude further screening in a second tier. One must also consider the time and effort involved in visual examination of hundreds of drops.

3.2. IMPORTANCE OF THE FREEZING PROTOCOL

Given that the initial screen does not usually produce X-ray ready crystals, optimization of the promising hits is the second step. Even if the initial crystals do exhibit excellent diffraction quality, a drug-discovery program needs to produce more of them for further experiments with the inhibitors. This requires a reproducible and steady supply of the crystals. Batch to batch differences in protein production can lead to irreproducibility in the crystallization, which is why it is clearly an advantage to repeat the crystallization with one and the same batch. At the same time, storage of the batch introduces variations because the protein ages with time. To improve reproducible outcomes from stored protein batches, we use a method which involves flash-freezing the protein solution in aliquots of less than 100 μL in thin-walled Eppendorf tubes for storage at –70°C and then rapid thawing at 37°C.[4]

3.3. THE ROLE OF THE PROTEIN BUFER BEFORE CRYSTALLIZATION SCREENING

Nucleation occurs at high levels of supersaturation. The more protein molecules that are in solution, the more likely it is that a critical mass is reached which can lead to a stable nucleus upon which further growth can occur. If the protein is poorly soluble in a particular buffer or pH, it may never reach a high enough level of supersaturation to support a nucleation event. Thus a higher, rather than lower, protein concentration in the crystallization screening is advantageous. The buffer choice can be critical, but it is often not optimized after the last purification step. Instead the buffer used in the elution of the last chromatographic column becomes the buffer by default in which the protein is concentrated for the crystallization trials. For example, we had one protein that would not concentrate to more than 0.1 mg/mL in the SEC buffer of 10 mM Tris-HCl, 150 mM NaCl, pH 8.0. The protein could be concentrated to 10 mg/mL after exchanging the SEC buffer for a phosphate buffer at the same pH of 8.0. In another case, a protein that precipitated heavily after a few hours in the SEC buffer, crystallized in one of one of the screen buffers without any precipitant. The protein solubility as a function of buffer/pH is easy to test and can therefore be done at an early level of the screening. It is especially useful to examine when the protein does not concentrate to more than 1–2 mg/mL in the SEC buffer.

4. Summary

This chapter covers some of the tenets of the crystallization approach used by our academic laboratory. It is a small laboratory with no automation except a crystallization robot. The suggestions here are not used to the exclusion of the many other options available, such as Thermofluor stability studies, dynamic light scattering, modification of the surface entropy, domain refinement, etc. We use these and other methods when the first screens fail. I have chosen to focus on the ones that I have in this chapter because they are simple to implement. For that reason, they should be considered as a first recourse. For example, changing the buffer of the protein is certainly easier and quicker than cloning a new construct of it.

The size and type of laboratory dictates what approaches are practical, cost-effective, and efficient. The approaches presented here have met these three criteria in our laboratory and they have proved successful. Our experiences may be relevant for other academic laboratories or drug-discovery programs.

References

1. J. Jancarik and S.H. Kim, *Acta Crystallographica* D60, 1670–1673 (2004).
2. A. Izaac, C. Schall and T. Mueser, *Acta Crystallographica* D62, 833–842 (2006).
3. B.K. Collins, R. Stevens and R. Page, *Acta Crystallographica* F61, 1035–1038 (2005).
4. J. Deng, D.R. Davies, G. Wisedchaisri, M. Wu, W. Hol and C. Mehlin, *Acta Crystallographica* D60, 203–204 (2004).
5. M. Harris and T.A. Jones, *Acta Crystallographica* D58, 1889–1891 (2002).
6. A. D'Arcy, F. Villard and M. March, *Acta Crystallographica* D63, 550–554 (2007).
7. G.C. Ireton and B.L. Stoddard, *Acta Crystallographica* D60, 601–605 (2003).
8. M. Carson, D.H. Johnson, H. McDonald, C. Brouillette and L.J. DeLucas, *Acta Crystallographica* D63, 295–301.
9. A. Hassell, G. An, R.K. Bledsoe, J.M. Bynum, H.L. Carter III, S-J. J. Deng, R.T. Gampe, G.T.E. Grisard, K.P. Madauss, R.T. Nolte, W.J. Rocque, L. Wang, K.L. Weaver, S.P. Williams, G.B. Wisely, R. Xu and L.M. Shewchuk, *Acta Crystallographica* D63, 72–79 (2007).
10. B.W. Segelke, *Journal of Crystal Growth*, 232, 553–562 (2001).
11. J. Newman, D. Egan, T.S. Walter, R. Meged, I. Berry, M. Ben Jelloul, J.L. Sussman, D.I. Stuart and A. Perrakis, *Acta Crystallographica* D61, 1426–1431 (2005).

FRAGMENT-BASED DRUG DISCOVERY IN ACADEMIA: EXPERIENCES FROM A TUBERCULOSIS PROGRAMME

TIMO J. HEIKKILA[1], SACHIN SURADE[1], HERNANI L. SILVESTRE[1], MARCIO V.B. DIAS[1], ALESSIO CIULLI[2], KAREN BROMFIELD[2], DUNCAN SCOTT[2], NIGEL HOWARD[2], SHIJUN WEN[2], ALVIN HUNG WEI[2], DAVID OSBORNE[2], CHRIS ABELL[2], TOM L. BLUNDELL[1*]

[1]*Department of Biochemistry, University of Cambridge, Tennis Court Road, Cambridge CB2 1GA, United Kingdom*

[2]*University Chemical Laboratory, Lensfield Road, Cambridge CB2 1EW, UK*

Abstract. The problems associated with neglected diseases are often compounded by increasing incidence of antibiotic resistance. Patient negligence and abuse of antibiotics has lead to explosive growth in cases of tuberculosis, with some *M. tuberculosis* strains becoming virtually untreatable. Structure-based drug development is viewed as cost-effective and time-consuming method for discovery and development of hits to lead compounds. In this review we will discuss the suitability of fragment-based methods for developing new chemotherapeutics against neglected diseases, providing examples from our tuberculosis programme.

Keywords: Fragment-based; drug discovery; tuberculosis; resistance

1. Introduction

Tuberculosis (TB) remains one of the deadliest diseases on the planet, claiming the lives of approximately two million people each year (WHO, 2008). Moreover, the mortality rates are once again on the rise. This is attributed to the emergence of multi-drug resistant (MDR) strains, and more

* To whom correspondence should be addressed. Tom L. Blundell, Department of Biochemistry, University of Cambridge, Tennis Court Road, Cambridge CB2 1GA, United Kingdom; e-mail: tom@cryst.bioc.cam.ac.uk

recently, extensively drug-resistant (XDR) strains of *Mycobacterium tuberculosis*. Strains resistant to isoniazid and rifampicin, important components in the first-line of drug treatment, are categorised as MDR, while XDR strains are defined as those that are also resistant to at least three of the six classes of second-line drugs, seriously limiting treatment options and making XDR-TB virtually untreatable. Furthermore, the global HIV epidemic has produced a new and highly susceptible population, and this has increased incidence of TB as well as TB-related deaths in many parts of the world. The investment in development of new antibacterials has waned over recent decades and there has been an inadequate response to the resurgence of TB.

One of the major challenges in treatment of TB is the ability of *M. tuberculosis* to switch into a dormant, latent lifestyle upon gaining entry to pulmonary macrophages. The organism undergoes a metabolic shutdown and consequently many of the protein targets for antibiotics, such as the translational machinery of the cell, only operate at a basal level in this state (Tufariello et al., 2003; Cardona, 2007). This means that during the dormant phase the bacilli are particularly difficult to kill and as a consequence of this persistence, drug treatment has to be extended. Most of the current TB drugs require long courses of treatment in order to completely clear the patients of *M. tuberculosis* and prevent relapse. Currently, even the most effective regimens require a combination of at least three drugs and last for 6 months (WHO, 2008). As patients often start to feel better within a few weeks, they have little motivation to complete therapy and frequently stop taking the antibiotics. The latent, persistent bacilli are not completely cleared by such short courses of antibiotics, and this has directly contributed to the emergence of drug-resistant TB strains. To address this, current WHO guidelines call for treatment to be directly observed (DOTS scheme). One other significant challenge is the lack of infrastructure for drug delivery and treatment supervision, particularly in areas that are afflicted by poverty and unstable governments. Unfortunately, these are also often the areas worst affected by the disease.

In the face of a rapidly deteriorating situation and relative lack of interest from industry (with notable exceptions), academic research groups must take more responsibility for identifying novel drug targets as well as for early-stage discovery of novel antitubercular agents. Consequently, new approaches are important for target identification and validation as well as lead discovery. In this review we discuss the new technologies available for target identification and assess the suitability of fragment-based lead discovery and optimisation for addressing the issue of early-stage drug development in academia, illustrating different stages of the process with results from our TB programme.

2. Tuberculosis: target identification and promising drug targets

Although the complete genome of *M. tuberculosis* became available in 1998 (Cole et al., 1998) and provided unprecedented opportunities for target-specific drug discovery, progress has been slow. This is mainly due to a lack of a sufficiently strong interest by the pharmaceutical and biotechnology industries. With resistance emerging for many of the most commonly used TB drugs, there is a constant need for new targets for drug discovery. Along with more traditional experimental approaches, computational studies can also contribute to drug target identification. One attractive approach to target identification in sequenced genomes is based on phylogenetic tree analysis of proteins (Searls, 2003; Liao et al., 2008). A more recent method is based on systems biology approaches where interdependent biochemical pathways are studied simultaneously. This systems biology approach can yield important information and recently, a server based on systems approach has been set up for *M. tuberculosis* (Beste et al., 1996).

Candidate drug leads for potential targets can be discovered using high-throughput or fragment-based screening and optimised using structure-based drug design. Access to well-diffracting crystals is one of the prerequisites for successful application of these techniques, although nuclear magnetic resonance offers an alternative structure-based approach. Structure-based virtual screening and other computational approaches can also contribute (Kairys et al., 2006; Radestock et al., 2008) and comparative models for various *M. tuberculosis* proteins are available (Silveira et al., 2005, 2006). Recently, our group has carried out a structural analysis of nsSNPs and their effects on protein structure and interactions in an attempt to correlate this with disease (Worth et al., 2008). While similar information is not yet available for drug-resistant strains of *M. tuberculosis*, it would ultimately allow correlation of the resistance-causing mutations with three-dimensional structures of proteins. This could throw light on the mechanisms of resistance and stimulate ideas on how they might be overcome. It could also reveal easily mutatable drug targets, thus helping to make better early decisions on commitment of resources on potential targets.

Structure-based drug discovery is widely seen as one of the most promising techniques for addressing early-stage drug development for neglected diseases (Blundell, 1996; Sorensen et al., 2006; Holton et al., 2007; Congreve et al., 2005) The structures of a number of potential TB drug targets have already been solved and can readily be used for structure-based techniques, such as virtual high-throughput screening and fragment-based approaches. Some interesting targets are described below. Each represents a different paradigm, but brings with it its own set of challenges.

One of the more obvious TB drug targets is the unique cell envelope of *M. tuberculosis* that differs substantially from the cell wall structures of both Gram-negative and Gram-positive bacteria. This cell wall composition accounts for its unusual low permeability and resistance towards common antibiotics (Dover et al., 2004). The main structural element consists of a cross-linked network of peptidoglycan in which some of the muramic acid residues are replaced with a complex polysaccharide, arabinogalactan. The arabinogalactan is attached to peptidoglycan through a unique linker unit, and in turn is acylated at its distal end to peptidoglycan with mycolic acids. The entire complex, mycolylarabinogalactan–peptidoglycan or mAGP, is essential for viability in *M. tuberculosis* and other mycobacteria (Dover et al., 2004).

Of the components of the cell wall, the mycolic acids are perhaps the most interesting. These long chain, α-alkyl, β-hydroxyl fatty acids give rise to important characteristics of the organism, including resistance to chemical injury and dehydration, low permeability to antibiotics, virulence, acid-fast staining and the ability to persist within the host (Barry et al., 1998; Dubnau et al., 2000). The synthesis of mycolic acids is the target of front-line anti-tubercular drugs isoniazid and ethambutol (Tonge et al., 2007). Furthermore, cyclopropanation of mycolic acids has been shown to have a profound effect on the resistance of the mycobacteria to the oxidative stress and the fluidity and permeability of the cell wall (George et al., 1995; Huang et al., 2002). Consequently, cyclopropane synthetases required for this process are considered as good targets against persistent TB, and structures of three of these enzymes have been determined (Huang et al., 2002).

The biosynthetic pathways leading to formation of key mycobacterial cell wall components are similarly attractive targets for the rational design of new antituberculosis agents. The phospholipids present in mycobacterial cell envelopes are almost invariably derivatives of phosphatidic acid. The most common are the phosphatidylinositol mannosides (PIMs) and higher order glycolipids and lipoglycans such as lipomannan (LM) and lipoarabinomannan (LAM), which all play key roles in mycobacterial physiology. Genome sequencing together with genetic manipulation of mycobacteria has led to the identification of some of the enzymes involved in the early stages of PIM, LM, and LAM biosynthesis. The phosphatidyl-myo-inositol mannosyltransferase (PimA, E.C. 2.4.1.57) catalyses the condensation of the first mannosyl residue to phosphatidylinositol using GDP-Mannose as a cofactor, yielding phosphatidylinositol monomannoside (PIM1). This enzyme appears to be essential for mycobacterial growth and no human homologues have been identified (Kordulàková et al., 2003).

The crystal structures of PimA in complex with GDP and GDP-Man show a two-domain organisation typical of GT-B glycosyltransferases, and this has led to the proposal of a significant hinge bending motions between the two domains during catalysis (Guerin et al., 2007). The high affinity of GDP/GDP-Man (K_D ~10^{-7} M) and the nature of the active site cleft point to a potential good druggabililty of PimA. However, the unavailability of a crystal structure of the apoenzyme, the absence of known inhibitors and the difficulty in assaying the activity of the enzyme all provide challenges to rational drug design.

Several other pathways are shared between various bacterial species but are not found in humans, thus making them obvious targets for drug development. One such example is the shikimate pathway which facilitates the biosynthesis of aromatic rings from carbohydrate precursors in micro-organisms and plants. The shikimate pathway has been found essential in algae, bacteria, and fungi, but it is lacking in mammals, thus necessitating salvage of aromatic compounds from food (Bentley, 1990). The pathway consists of seven steps, starting from phosphoenolpyruvate (PEP) and D-erythrose 4-phosphate (E4P) and ultimately producing the branch point compound chorismate. This is then utilised for several additional terminal pathways. Structures of many of the enzymes of the shikimate pathway from *M. tuberculosis* are available, including shikimate kinase (aroK), 3-dehydroquinate dehydratase (aroD), and EPSP synthase (Gourley et al., 1999; Dias et al., 2007), thus making them attractive targets for drug discovery projects.

The *aroK* gene encoding shikimate kinase has been shown to be essential for the survival of *M. tuberculosis* (Parish and Stoker, 2002). It has been the focus of several high throughput screening projects in industry; however, no strong lead compounds have surfaced. While numerous crystal structures of *M. tuberculosis* shikimate kinase are available (19 PDB entries to date), there are several challenges to overcome when considering shikimate kinase as a target for structure-based drug design. It is apparent that the enzyme undergoes large conformational changes between open and closed structures upon substrate binding (Hartmann et al., 2006). Furthermore the active site, even in the closed structures, is relatively solvent exposed and the protein exhibits a high pI which is thought to contribute to the enzymes promiscuous inhibition and sensitivity to salt concentration (Dias et al., 2007).

The gene *aroD*, formally named *aroQ* in *M. tuberculosis* (Garbe et al., 1991), encodes the type II 3-dehydroquinate dehydratase. Whilst a number of potent inhibitors of the enzyme have been described (González-Bello and Castedo, 2007; Toscano et al., 2007), they are not ideal for further drug development due to the difficult chemistry involved in their synthesis. A number of crystal structures of the enzyme have been published, both *apo*-form, and with inhibitors bound. These structures assist in the identification of key interactions between inhibitor and protein, but also reveal two problems. Firstly, the enzyme is a dodecamer making structural studies more challenging. Secondly, the presence of a highly mobile loop containing key catalytic residues complicates structure-based inhibitor design, since this loop appears to be engaged in the closed crystal structures with inhibitors bound but is disordered in the more open *apo* structures.

Another promising pathway to target is the glyoxylate shunt. It has been shown to play a crucial role in the survival of persistent *M. tuberculosis* and thought not to operate in humans, therefore providing further targets for development of new antitubercular agents active against the latent form of the disease. The strategy for survival of TB during chronic stages of infection is thought to involve a metabolic shift in the bacteria's carbon source to C_2 substrates generated by the β-oxidation of fatty acids. Under these conditions, glycolysis is decreased and the glyoxylate shunt is significantly upregulated allowing anaplerotic maintenance of the tricarboxylic acid (TCA) cycle (McKinney et al., 2000). The glyoxylate shunt converts isocitrate to succinate and glyoxylate, catalysed by the enzyme isocitrate lyase 1 (ICL1), followed by the addition of acetyl-CoA to glyoxylate to form malate by malate synthase (Sharma et al., 2000; Smith et al., 2003). It has been shown that expression of ICL1 is upregulated under certain growth conditions and during infection of macrophages (McKinney et al., 2000). Furthermore, ICL1 is required for the survival of bacteria in activated macrophages but not in resting macrophages (McKinney et al., 2000). It has also been demonstrated that ICL1 is important for survival of *M. tuberculosis* in the lungs of mice during the persistent phase of infection, but is not essential during the acute phase of infection (McKinney et al., 2000).

Finally one could target the synthesis of pantothenate, or vitamin B5. Pantothenate synthase is the third enzyme of the pathway in bacteria, which is essential not only for pantothenate but also coenzyme A biosynthesis. It catalyses the condensation of pantoate and ATP, with the subsequent hydrolysis and release of pyrophosphate, followed by condensation of the resulting pantoyladenylate intermediate with β-alanine. A strain of *Mycobacterium tuberculosis* with a pantothenate synthetase knock-out is severely attenuated in mice, thus making this enzyme an attractive drug target (Sambandamurthy et al., 2002). The crystal structure of pantothenate synthetase has been determined to high resolution in *E. coli* (von Delft et al., 2001) as well as in *M. tuberculosis* in complex with substrate and product small-molecule ligands (Wang and Eisenberg, 2006) and in complex with potent inhibitors that mimic the structure of the reaction intermediate (Ciulli et al., 2008). These crystal structures confirm pantothenate synthetase as a member of the cytidyltransferase superfamily, suggesting particular lines of approach for structure-based strategy for drug discovery. Current research on this target is very active, with a number of promising inhibitors already identified from high-throughput screening programmes (Velaparthi et al., 2008; White et al., 2007). We have successfully conducted fragment screening against pantothenate synthetase, as described in Section 3.3.

3. Fragment-based drug development

The fragment-based drug development approach is based on the premise that fragment-like molecules, owing to their small sizes, are more likely to bind specifically to proteins than larger, drug-like compounds, albeit with a much weaker affinity. Furthermore, fragment screening allows exploration of much larger chemical space than traditional high-throughput screening of drug-like molecules. Although the size and the content of the fragment library must limit the chemical space explored by the screening exercise, well constructed libraries (Congreve et al., 2003) of a thousand or less fragments have proved successful with a wide range of targets.

Fragment-based approaches have become a standard drug development method in industry, with a number of pharmaceutical companies relying on this technique to produce novel lead compounds, even against targets previously found difficult to inhibit (Alex and Flocco, 2007). With commercially available fragment screening libraries and more cost-effective screening methods becoming available, academic groups have also started to apply various fragment-based techniques to identify hits and develop new lead molecules (Bosch et al., 2006; Caldwell et al., 2008). The fragment-based drug development process can be split in three distinct steps; fragment screening, fragment hit validation and fragment growing or linking. These are discussed in detail in the following paragraphs.

3.1. FRAGMENT SCREENING

A number of biophysical techniques can be used for the initial screening of fragments. One of the simplest assays is the thermofluor-based thermal shift experiment, in which compounds are added to the target in the presence of a fluorescent dye that binds preferentially to the unfolded state of a protein (Lo et al., 2004). The samples are gradually heated in a real-time PCR machine and the fluorescence is monitored continuously. Hits are identified as compounds that stabilise the folded state of the target protein (Fig. 1) (Gould et al., 2006). Only relatively small shifts are generally seen from weakly-bound fragments, given that a correlation between the shift in unfolding temperature and the binding affinity is often observed (Lo et al., 2004). This technique is particularly useful for an academic fragment screening programme as it is both inexpensive to run and readily applicable in high-throughput manner.

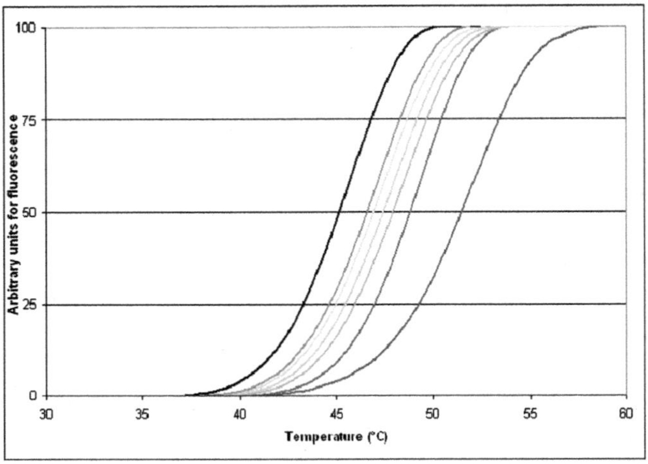

Figure 1. Typical results from thermal shift assays for fragment binding. Midpoint for unfolding of *M. tuberculosis* isocitrate lyase is at 45°C (black trace), with product glyoxylate stabilising the enzyme to produce a positive shift of 6.5°C (red trace). Fragment MB1 produces a positive shift of 3.5°C (magenta trace), while other fragments produce less significant shifts (pink, yellow, cyan and green traces).

Ligand-based NMR methods are also well-suited for fragment screening. These techniques monitor the resonances of the small molecules directly and are, consequently, not limited by the size of the target protein or prior knowledge of the protein NMR spectrum. One of the most useful techniques for ligand screening is WaterLOGSY (Lepre et al., 2004; Dalvit et al., 2001). This experiment detects fragment binding by magnetisation transfer from bulk water to fragments, via stably bound water molecules in the protein-fragment complex. A related technique is saturation transfer difference (STD), which exploits a magnetisation transfer process directly from the protein to the bound fragment (Mayer and Meyer, 2001). With these techniques, hits can be rapidly identified from cocktails of 3–4 fragments to minimise the overlap of signals. These binding assays can be followed by competition experiments with known ligands to determine the binding site as well as to eliminate interference from non-specific binding, a common caveat of highly sensitive NMR detection (Ciulli et al., 2006).

X-ray crystallography can also be used for fragment screening, although the throughput of this method is heavily influenced by access to synchrotron beamtime or a powerful home source, and is often not an option in academia. Nevertheless, this method can be very powerful as it provides direct validation of the binding of the fragment. Crystals can be soaked with cocktails of fragments at high concentration (up to 200 mM per compound; Hartshorn et al., 2005; Blundell et al., 2002). After data collection and processing, difference electron density maps are analysed to detect fragment binding. Depending on the quality of the data it might be necessary to break down the cocktail and to repeat the soaking experiment with single compounds to confirm the identity of the binder. Ideally, the hit rate should be less than one compound per cocktail in order to avoid multiple cases of partial occupancies, which would make fragment identification difficult.

3.2. HIT VALIDATION

Once hits are identified their binding needs to be confirmed in terms of affinity and mode. The most commonly used technique to quantify binding affinity is isothermal titration calorimetry (ITC; Fig. 2). ITC can provide information on the binding affinity, stoichiometry as well as the enthalpic and entropic contribution to the free energy of binding. Monitoring ΔH and ΔS by ITC may reveal changes in binding modes (Ciulli et al., 2006; Holdgate and Ward, 2005). Determining the exact binding affinity also allows calculation of ligand efficiency (binding energy divided by the number of non-hydrogen atoms), a useful metric that can be used to guide fragment selection and lead optimisation during the discovery process (Hopkins et al., 2004).

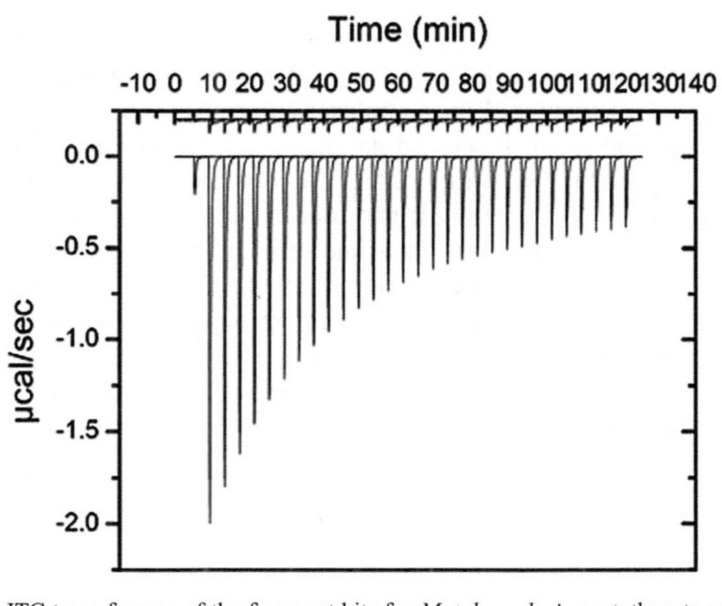

Figure 2. ITC trace for one of the fragment hits for *M. tuberculosis* pantothenate synthetase identified in thermal shift assays. K_D for the fragment was found to be 0.8 mM, with ΔG = -4.2 kcal/mol, ΔH = -9.5 kcal/mol and ΔS = -17.8 cal/mol^{-1}K^{-1}.

Ultimately, the binding mode and/or the three-dimensional structure of the fragment bound to the protein should be determined, ideally by X-ray crystallography. This will allow assessment of structure-activity relationship and is an important requirement before the compound is progressed into chemical optimisation. It should be noted however that in the absence of crystallographic information, competitive NMR experiments can be a useful tool to identify the binding site and gain some information with respect to the compounds binding mode and affinity, especially if the crystallography is seen as the bottleneck of the project in question (Chung, 2007). This concept is exemplified by the cases of PimA and shikimate kinase, as previously discussed, where obtaining a soakable crystal form is difficult and synthetic exploration has been driven by NMR data.

3.3. A CASE STUDY ON FRAGMENT SCREENING: TARGETING PANTOTHENATE SYNTHETASE

Pantothenate synthetase from *M. tuberculosis* was screened in a thermal denaturation assay against a fragment library of 1,300 compounds. Fragments which caused a positive shift of the protein's thermal melting temperature greater than 0.5°C were considered as hits, giving 23 fragments (hit rate ~2%). These hits were taken forward for further validation by ITC and NMR spectroscopy experiments. Compounds that were identified as ligands were then soaked into protein crystals of M. tuberculosis pantothenate synthetase and the structure of the complexes solved by X-ray crystallography. As an illustration of this process, a fragment containing a benzodioxole core displayed a good thermal shift of 2.5°C. This compound was titrated against the protein in an ITC experiment and found to have a KD of 1.2 mM, corresponding to a reasonably good ligand efficiency of 0.29. WaterLOGSY and STD experiments showed ligand binding and displacement by ATP. Finally the crystal structure of this fragment binding at the active site of pantothenate synthetase was determined (Fig. 3). This information provides a useful starting point for developing more potent inhibitors using structure-guided chemical synthesis.

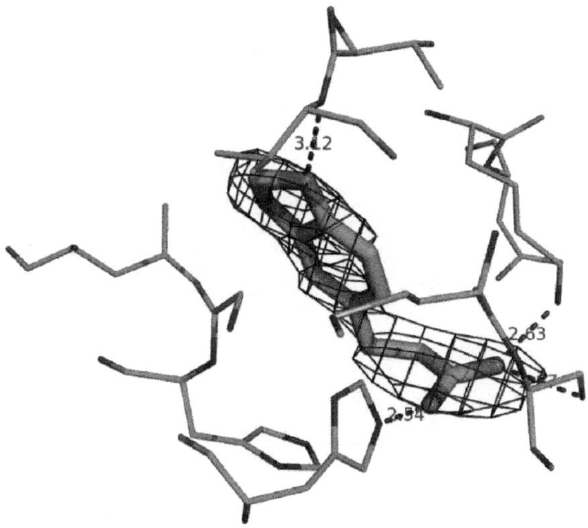

Figure 3. X-ray crystal structure of a fragment seen bound at the active site of *M. tuberculosis* pantothenate syntethase. The initial unbiased omit Fo-Fc electron density map is contoured around the fragment at 3.0 σ. Key hydrogen bond interactions and distances between the fragment and residues in the enzyme active site are shown in purple. The fragment had a thermal shift of 2.5°C and its binding was validated by STD and WaterLOGSY NMR spectroscopy experiments. The K_D of the fragment was found to be 1.2 mM from ITC measurements.

3.4. FRAGMENT GROWING AND LINKING

After hit validation, the aim is to elaborate the fragment hit to improve the binding affinity, ideally as an iterative process guided by structural information. There are two routes that could be taken, namely fragment growing and fragment linking (Fig. 4; Howard et al., 2006). The former involves chemical elaboration around a single fragment hit in order to improve binding by picking up new interactions within the target cavity. The latter approach requires two or more fragments that are found binding to different but adjacent sites within the active site of the target protein, and relies on the design of a chemical scaffold to combine these and improve the binding affinity by synergy. The linking strategy is, however, quite challenging in practise as the process is sensitive to designing a linker which does not perturb the binding mode of the original fragments (Hajduk and Greer, 2007). Both these strategies can be guided by using computational docking tools such Genetic Optimisation Ligand Docking (GOLD; Jones et al., 1995, 1997).

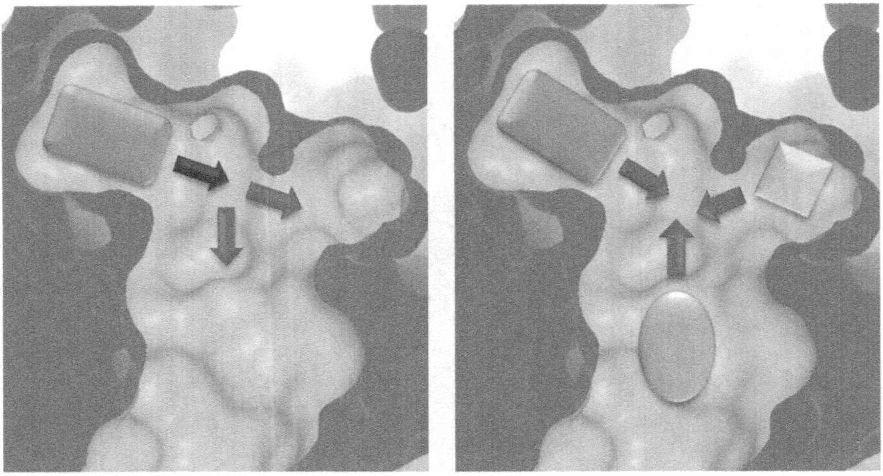

Figure 4. Diagram displaying the two hit elaboration strategies; (a) fragment growing and (b) fragment linking (reproduced with permission from Howard et al., 2006).

Ideally, in a fragment-based drug discovery process, a series of X-ray crystal structures of fragments are obtained. From this set, fragments are chosen that have good ligand efficiency, are synthetically accessible and possess suitable vectors from which it is possible to chemically elaborate. Selected fragments are then systematically elaborated to maximise favourable binding interactions between the ligand and the residues of the active site. The chemistry typically involves relatively high yielding reactions, e.g. amide bond formation, arylation and alkylation, reductive amination, and click chemistry. As a fragment is grown into an active site, there is the opportunity to form interactions with protein backbone residues and side-chains, in addition to complementing ligand shape with pockets within the active site. This can be done for example by the use of sulfonamides or small heterocyclic rings as linkers and potential hydrogen-bond donors/ acceptors. Throughout this growth process, it is desirable to optimise the number of rotatable bonds, to allow a certain degree of flexibility for the ligand to adopt the required binding pose whilst minimising the entropic penalty of binding. Finally, a drug-like compound would be expected to obey Lipinski's rule of five (Lipinski et al., 2001), and this guideline can be monitored and adhered to throughout the fragment development process.

4. Conclusions

The global tuberculosis epidemic is spiralling out of control due to drug-resistant strains of *M. tuberculosis*, and innovative solutions are needed for establishing new drug targets as well as for hit identification and lead optimisation. New computational approaches are already offering complementary methods to more traditional, experimental techniques in target identification and validation, with various algorithms being used to mine the genomic sequence of *M. tuberculosis*. The development and application of various structure-based techniques will certainly play a key role. Fragment-based drug development approach is considered as one of the most promising new methods for identifying new "hits" which can be grown into potential lead molecules. We have already achieved promising results using the fragment-based approach against TB targets as described in this review. Most of the techniques described are inexpensive and straightforward to perform, and similar programmes could be readily carried out in most academic labs.

References

Alex AA, Flocco MM. Fragment-based drug discovery: what has it achieved so far? *Curr Top Med Chem.* 2007; 7(16):1544–67.

Barry CE 3rd, Lee RE, Mdluli K, Sampson AE, Schroeder BG, Slayden RA, Yuan Y. Mycolic acids: structure, biosynthesis and physiological functions. *Prog Lipid Res.* 1998; 37(2–3):143–79.

Bentley R. The shikimate pathway–a metabolic tree with many branches. *Crit Rev Biochem Mol Biol.* 1990; 25(5):307–84.

Beste DJ, Hooper T, Stewart G, Bonde B, Avignone-Rossa C, Bushell ME, Wheeler P, Klamt S, Kierzek AM, McFadden J. GSMN-TB: a web-based genome-scale network model of *Mycobacterium tuberculosis* metabolism. *Genome Biol.* 2007; 8(5):R89.

Blundell TL. Structure-based drug design. *Nature.* 1996; 384(6604 Suppl):23–6.

Blundell TL, Jhoti H, Abell C. High-throughput crystallography for lead discovery in drug design. *Nat Rev Drug Discov.* 2002; 1(1):45–54.

Bosch J, Robien MA, Mehlin C, Boni E, Riechers A, Buckner FS, Van Voorhis WC, Myler PJ, Worthey EA, DeTitta G, Luft JR, Lauricella A, Gulde S, Anderson LA, Kalyuzhniy O, Neely HM, Ross J, Earnest TN, Soltis M, Schoenfeld L, Zucker F, Merritt EA, Fan E, Verlinde CL, Hol WG. Using fragment cocktail crystallography to assist inhibitor design of *Trypanosoma brucei* nucleoside 2-deoxyribosyltransferase. *J Med Chem.* 2006; 49(20):5939–46.

Caldwell JJ, Davies TG, Donald A, McHardy T, Rowlands MG, Aherne GW, Hunter LK, Taylor K, Ruddle R, Raynaud FI, Verdonk M, Workman P, Garrett MD, Collins I. Identification of 4-(4-Aminopiperidin-1-yl)-7H-pyrrolo[2,3-d]pyrimidines as Selective Inhibitors of Protein Kinase B through Fragment Elaboration. *J Med Chem.* 2008; 51(7):2147–57.

Cardona PJ. New insights on the nature of latent tuberculosis infection and its treatment. *Inflamm Allergy Drug Targets.* 2007; 6(1):27–39.

Chung CW. The use of biophysical methods increases success in obtaining liganded crystal structures. *Acta Crystallogr D Biol Crystallogr.* 2007; 63(Pt 1):62–71.

Ciulli A, Williams G, Smith AG, Blundell TL, Abell C. Probing hot spots at protein-ligand binding sites: a fragment-based approach using biophysical methods. *J Med Chem.* 2006; 49(16):4992–5000.

Ciulli A, Scott DE, Ando M, Reyes F, Saldanha SA, Tuck KL, Chirgadze DY, Blundell TL, and Abell C. Inhibition of *Mycobacterium tuberculosis* pantothenate synthetase by analogues of the reaction intermediate. *ChemBioChem.* 2008; 9(16):2606–2611.

Cole ST, Brosch R, Parkhill J, Garnier T, Churcher C, Harris D, Gordon SV, Eiglmeier K, Gas S, Barry CE 3rd, Tekaia F, Badcock K, Basham D, Brown D, Chillingworth T, Connor R, Davies R, Devlin K, Feltwell T, Gentles S, Hamlin N, Holroyd S, Hornsby T, Jagels K, Krogh A, McLean J, Moule S, Murphy L, Oliver K, Osborne J, Quail MA, Rajandream MA, Rogers J, Rutter S, Seeger K, Skelton J, Squares R, Squares S, Sulston JE, Taylor K, Whitehead S, Barrell BG. Deciphering the biology of *Mycobacterium tuberculosis* from the complete genome sequence. *Nature.* 1998; 393(6685):537–44.

Congreve M, Carr R, Murray C, Jhoti H. A 'rule of three' for fragment-based lead discovery? *Drug Discov Today.* 2003; 8(19):876–7.

Congreve M, Murray CW, Blundell TL. Structural biology and drug discovery. *Drug Discov Today.* 2005; 10(13):895–907.

Dalvit C, Fogliatto G, Stewart A, Veronesi M, Stockman B. WaterLOGSY as a method for primary NMR screening: practical aspects and range of applicability. *J Biomol NMR.* 2001; 21(4):349–59.

Dias MV, Faím LM, Vasconcelos IB, de Oliveira JS, Basso LA, Santos DS, de Azevedo WF Jr. Effects of the magnesium and chloride ions and shikimate on the structure of shikimate kinase from *Mycobacterium tuberculosis. Acta Crystallogr Sect F Struct Biol Cryst Commun.* 2007; 63(Pt 1):1–6.

Dover LG, Cerdeño-Tárraga AM, Pallen MJ, Parkhill J, Besra GS. Comparative cell wall core biosynthesis in the mycolated pathogens, *Mycobacterium tuberculosis* and Corynebacterium diphtheriae. *FEMS Microbiol Rev.* 2004; 28(2):225–50.

Dubnau E, Chan J, Raynaud C, Mohan VP, Lanéelle MA, Yu K, Quémard A, Smith I, Daffé M. Oxygenated mycolic acids are necessary for virulence of *Mycobacterium tuberculosis* in mice. *Mol Microbiol.* 2000; 36(3):630–7.

Garbe T, Servos S, Hawkins A, Dimitriadis G, Young D, Dougan G, Charles I. The *Mycobacterium tuberculosis* shikimate pathway genes: evolutionary relationship between biosynthetic and catabolic 3-dehydroquinases. *Mol Gen Genet.* 1991; 228(3):385–92.

George KM, Yuan Y, Sherman DR, Barry CE 3rd. The biosynthesis of cyclopropanated mycolic acids in *Mycobacterium tuberculosis*. Identification and functional analysis of CMAS-2. *J Biol Chem.* 1995; 270(45):27292–8.

González-Bello C, Castedo L. Progress in type II dehydroquinase inhibitors: from concept to practice. *Med Res Rev.* 2007; 27(2):177–208.

Gould TA, van de Langemheen H, Muñoz-Elías EJ, McKinney JD, Sacchettini JC. Dual role of isocitrate lyase 1 in the glyoxylate and methylcitrate cycles in *Mycobacterium tuberculosis. Mol Microbiol.* 2006; 61(4):940–7.

Gourley DG, Shrive AK, Polikarpov I, Krell T, Coggins JR, Hawkins AR, Isaacs NW, Sawyer L. The two types of 3-dehydroquinase have distinct structures but catalyze the same overall reaction. *Nat Struct Biol.* 1999; 6(6):521–5.

Guerin ME, Kordulakova J, Schaeffer F, Svetlikova Z, Buschiazzo A, Giganti D, Gicquel B, Mikusova K, Jackson M, Alzari PM. Molecular recognition and interfacial catalysis by the essential phosphatidylinositol mannosyltransferase PimA from mycobacteria. *J Biol Chem.* 2007; 282(28):20705–14.

Hajduk PJ, Greer J. A decade of fragment-based drug design: strategic advances and lessons learned. *Nat Rev Drug Discov.* 2007; 6(3):211–9.

Hartmann MD, Bourenkov GP, Oberschall A, Strizhov N, Bartunik HD. Mechanism of phosphoryl transfer catalyzed by shikimate kinase from *Mycobacterium tuberculosis*. *J Mol Biol.* 2006; 364(3):411–23.

Hartshorn MJ, Murray CW, Cleasby A, Frederickson M, Tickle IJ, Jhoti H. Fragment-based lead discovery using X-ray crystallography. *J Med Chem.* 2005; 48(2):403–13.

Holdgate GA, Ward WH. Measurements of binding thermodynamics in drug discovery. *Drug Discov Today.* 2005; 10(22):1543–50.

Holton SJ, Weiss MS, Tucker PA, Wilmanns M. Structure-based approaches to drug discovery against tuberculosis. *Curr Protein Pept Sci.* 2007; 8(4):365–75.

Hopkins AL, Groom CR, Alex A. Ligand efficiency: a useful metric for lead selection. *Drug Discov Today.* 2004; 9(10):430–1.

Howard N, Abell C, Blakemore W, Chessari G, Congreve M, Howard S, Jhoti H, Murray CW, Seavers LC, van Montfort RL. Application of fragment screening and fragment linking to the discovery of novel thrombin inhibitors. *J Med Chem.* 2006; 49(4):1346–55.

Huang CC, Smith CV, Glickman MS, Jacobs WR Jr, Sacchettini JC. Crystal structures of mycolic acid cyclopropane synthases from *Mycobacterium tuberculosis*. *J Biol Chem.* 2002; 277(13):11559–69.

Jones G, Willet P, Glen RC. Molecular recognition of receptor sites using a genetic algorithm with a description of desolvation. *J Mol Biol.* 1995; 245(1): 43–53.

Jones G, Willet P, Glen RC, Leach AR, Taylor R. Development and validation of a genetic algorithm for flexible docking. *J Mol Biol.* 1997; 267(3): 727–48.

Kairys V, Fernandes MX, Gilson MK. Screening drug-like compounds by docking to homology models: a systematic study. *J Chem Inf Model.* 2006; 46(1):365–79.

Kordulákova J, Gilleron M, Puzo G, Brennan PJ, Gicquel B, Mikusová K, Jackson M. Identification of the required acyltransferase step in the biosynthesis of the phosphatidylinositol mannosides of mycobacterium species. *J Biol Chem.* 2003; 278(38):36285–95.

Lepre CA, Moore JM, Peng JW. Theory and applications of NMR-based screening in pharmaceutical research. *Chem Rev.* 2004; 104(8):3641–76.

Liao YL, Sun YM, Chau GY, Chau YP, Lai TC, Wang JL, Horng JT, Hsiao M, Tsou AP. Identification of SOX4 target genes using phylogenetic footprinting-based prediction from expression microarrays suggests that overexpression of SOX4 potentiates metastasis in hepatocellular carcinoma. *Oncogene.* 2008; in press.

Lipinski CA, Lombardo F, Dominy BW, Feeney PJ. Experimental and computational approaches to estimate solubility and permeability in drug discovery and development settings. *Adv Drug Deliv Rev.* 2001; 46(1–3):3–26.

Lo MC, Aulabaugh A, Jin G, Cowling R, Bard J, Malamas M, Ellestad G. Evaluation of fluorescence-based thermal shift assays for hit identification in drug discovery. *Anal Biochem.* 2004; 332(1):153–9.

Mayer M, Meyer B. Group epitope mapping by saturation transfer difference NMR to identify segments of a ligand in direct contact with a protein receptor. *J Am Chem Soc.* 2001; 123(25):6108–17.

McKinney JD, Höner zu Bentrup K, Muñoz-Elías EJ, Miczak A, Chen B, Chan WT, Swenson D, Sacchettini JC, Jacobs WR Jr, Russell DG. Persistence of *Mycobacterium tuberculosis* in macrophages and mice requires the glyoxylate shunt enzyme isocitrate lyase. *Nature.* 2000; 406(6797):735–8.

Parish T, Stoker NG. The common aromatic amino acid biosynthesis pathway is essential in *Mycobacterium tuberculosis*. *Microbiology.* 2002; 148(Pt 10):3069–77.

Radestock S, Weil T, Renner S. Homology model-based virtual screening for GPCR ligands using docking and target-biased scoring. *J Chem Inf Model.* 2008; 48(5):1104–17.

Sambandamurthy VK, Wang X, Chen B, Russell RG, Derrick S, Collins FM, Morris SL, Jacobs WR Jr. A pantothenate auxotroph of *Mycobacterium tuberculosis* is highly attenuated and protects mice against tuberculosis. *Nat Med.* 2002; 8(10):1171–4.

Sharma V, Sharma S, Hoener zu Bentrup K, McKinney JD, Russell DG, Jacobs WR Jr, Sacchettini JC. Structure of isocitrate lyase, a persistence factor of *Mycobacterium tuberculosis*. *Nat Struct Biol.* 2000; 7(8):663–8.

Silveira NJ, Bonalumi CE, Uchôa HB, Pereira JH, Canduri F, de Azevedo WF. DBMODELING: a database applied to the study of protein targets from genome projects. *Cell Biochem Biophys.* 2006; 44(3):366–74.

Silveira NJ, Uchôa HB, Pereira JH, Canduri F, Basso LA, Palma MS, Santos DS, de Azevedo WF Jr. Molecular models of protein targets from *Mycobacterium tuberculosis*. *J Mol Model.* 2005; 11(2):160–6.

Searls DB. Pharmacophylogenomics: genes, evolution and drug targets. *Nat Rev Drug Discov.* 2003; 2(8):613–23.

Smith CV, Huang CC, Miczak A, Russell DG, Sacchettini JC, Höner zu Bentrup K. Biochemical and structural studies of malate synthase from *Mycobacterium tuberculosis*. *J Biol Chem.* 2003; 17; 278(3):1735–43.

Sorensen TL, McAuley KE, Flaig R, Duke EM. New light for science: synchrotron radiation in structural medicine. *Trends Biotechnol.* 2006; 24(11):500–8.

Tonge PJ, Kisker C, Slayden RA. Development of modern InhA inhibitors to combat drug resistant strains of *Mycobacterium tuberculosis*. *Curr Top Med Chem.* 2007; 7(5):489–98.

Toscano MD, Payne RJ, Chiba A, Kerbarh O, Abell C. Nanomolar inhibition of type II dehydroquinase based on the enolate reaction mechanism. *ChemMedChem.* 2007; 2(1):101–12.

Tufariello JM, Chan J, Flynn JL. Latent tuberculosis: mechanisms of host and bacillus that contribute to persistent infection. *Lancet Infect Dis.* 2003; 3(9):578–90.

Velaparthi S, Brunsteiner M, Uddin R, Wan B, Franzblau SG, Petukhov PA. 5-tert-Butyl-N-pyrazol-4-yl-4,5,6,7-tetrahydrobenzo[d]isoxazole-3-carboxamide Derivatives as Novel Potent Inhibitors of *Mycobacterium tuberculosis* Pantothenate Synthetase: Initiating a Quest for New Antitubercular Drugs. *J Med Chem.* 2008; 51(7):1999–2002.

von Delft F, Lewendon A, Dhanaraj V, Blundell TL, Abell C, Smith AG. The crystal structure of *E. coli* pantothenate synthetase confirms it as a member of the cytidylyltransferase superfamily. *Structure.* 2001; 9(5):439–50.

Wang S, Eisenberg D. Crystal structure of the pantothenate synthetase from *Mycobacterium tuberculosis*, snapshots of the enzyme in action. *Biochemistry.* 2006; 45(6):1554–61.

White EL, Southworth K, Ross L, Cooley S, Gill RB, Sosa MI, Manouvakhova A, Rasmussen L, Goulding C, Eisenberg D, Fletcher TM 3rd. A novel inhibitor of *Mycobacterium tuberculosis* pantothenate synthetase. *J Biomol Screen.* 2007; 12(1):100–5.

WHO Report 2008, Global tuberculosis control - surveillance, planning, financing, WHO/HTM/TB/2008.393

Worth CL, Bickerton GR, Schreyer A, Forman JR, Cheng TM, Lee S, Gong S, Burke DF, Blundell TL. A structural bioinformatics approach to the analysis of nonsynonymous single nucleotide polymorphisms (nsSNPs) and their relation to disease., *J Bioinform Comput Biol.* 2007; 5(6):1297–318.

STRUCTURAL BIOLOGY CONTRIBUTIONS TO THE DISCOVERY OF DRUGS TO TREAT CHRONIC MYELOGENOUS LEUKEMIA

SANDRA W. COWAN-JACOB*, GABRIELE FENDRICH, ANDREAS FLOERSHEIMER, PASCAL FURET, JANIS LIEBETANZ, GABRIELE RUMMEL, PAUL RHEINBERGER, MARIO CENTELEGHE, DORIANO FABBRO, PAUL W. MANLEY

Novartis Institutes for Biomedical Research, CH-4056 Basel, Switzerland

Abstract. This case study illustrates how the determination of multiple co-crystal structures of the protein tyrosine kinase c-Abl was used to support drug discovery efforts leading to the design of nilotinib, a newly approved therapy for imatinib-intolerant and – resistant chronic myelogenous leukemia. Chronic myelogenous leukemia (CML) results from the BCR-Abl oncoprotein, which possesses a constitutively activated Abl tyrosine kinase domain. Although many chronic-phase CML patients treated with imatinib as first-line therapy maintain excellent, durable responses, patients who have progressed to advanced-stage CML frequently fail, or lose their response to therapy, often due to the emergence of drug-resistant mutants of the protein. More than 60 such point mutations have been detected in imatinib-resistant patients. We determined the crystal structures of wild-type and mutant Abl kinase in complex with imatinib and other small molecule Abl inhibitors, with the aim of understanding the molecular basis for resistance and to aid in the design and optimization of inhibitors active against the resistance mutants. These results are presented in a way which illustrates the approaches used to generate multiple structures, the type of information that can be gained and the way this information is used to support drug discovery.

Keywords: Tyrosine kinase, crystal structure, drug discovery, imatinib; nilotinib

* To whom correspondence should be addressed. Sandra Jacob, Novartis Institutes for Biomedical Research, Novartis Campus, Forum 1, CH-4056 Basel, Switzerland; e-mail: sandra.jacob@novartis.com

J.L. Sussman and P. Spadon (eds.), *From Molecules to Medicines*,

1. Introduction

Chronic myelogenous leukemia (CML) is caused by DNA damage leading to a gene defect in a hematological stem cell (HSC), resulting in the expression of the BCR-Abl oncoprotein (Melo and Barnes, 2007). In contrast to the tightly regulated c-Abl kinase, an auto-regulatory domain in the oncoprotein is truncated, leading to constitutive activation of the tyrosine kinase activity (Fig. 1). The resulting unregulated phosphorylation of intracellular proteins in HSCs leads to the uncontrolled growth and survival of the leukemic cells.

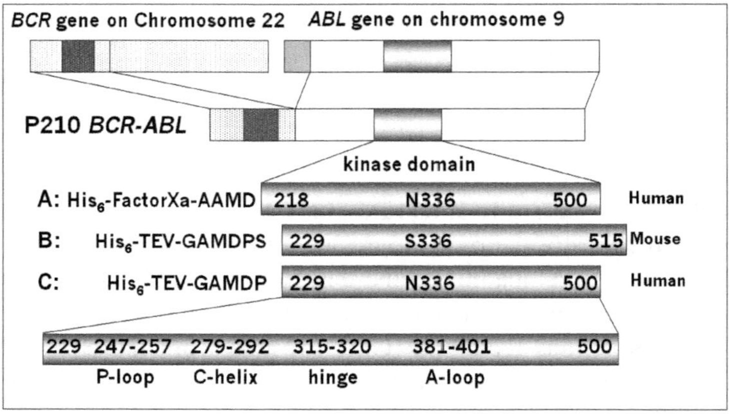

Figure 1. Schematic diagram showing the relationship between ABL/Abl and BCR-ABL/BCR-Abl (top), the constructs used for crystallography (middle) and the amino-acid numbers for structural elements mentioned in the text (bottom). The spotted dark segment shows the region of BCR-ABL that expresses the Abl-SH2-binding domain which must be phosphorylated for the activation of the catalytic region encoded by the ABL kinase domain (shaded bar). The flat grey segment shows the region of the ABL gene that is lost during the reciprocal translocation, and which is normally involved in the down regulation of the Abl kinase.

The therapeutic concept of BCR-Abl tyrosine kinase inhibition as a treatment modality for CML has been established with imatinib (Gleevec®; Novartis Pharma AG), which is an inhibitor of the Abl tyrosine kinase (Baccarani et al., 2006). In most chronic phase CML patients, imatinib therapy affords progression-free survival, with durable hematological or cytogenetic responses. However, a population of BCR-Abl expressing cells remain, such that the disease is not eradicated and a small percentage of patients develop insensitivity to imatinib and relapse (Druker, 2006). Relapse is frequently due to the expression of imatinib-resistant mutant forms of BCR-Abl, which escape inhibition through the exchange of amino-acid residues in the Abl kinase domain with alternative residues which maintain enzymatic activity, but have a reduced binding affinity to imatinib.

The good understanding of imatinib therapy has greatly benefited from X-ray crystallographic and NMR studies of the enzyme (Schindler et al., 2000; Nagar et al., 2002; Manley et al., 2002; Vajpai et al., 2008), which show that this drug inhibits the catalytic activity of BCR-Abl, by binding to an inactive conformation of the Abl kinase domain. Additional structural biology studies have established the mechanisms whereby mutant forms of the oncoprotein were resistant to imatinib (Gorre et al., 2001, Cowan-Jacob et al., 2004). This has paved the way for the design of second generation BCR-Abl inhibitors designed to inhibit the wild-type kinase and maintain activity against the imatinib-resistant mutants. It is believed that such new drugs will provide benefit for imatinib-resistant and – intolerant CML patients and, if used in combination with other distinct BCR-Abl inhibitors, might circumvent the emergence of drug-resistant mutant forms of BCR-Abl.

Crystallographic studies were undertaken in order to contribute to drug discovery efforts to find new compounds that might inhibit BCR-Abl with higher affinity while retaining the excellent kinase selectivity profile of imatinib. These studies included determining the binding modes of established chemotypes in order to understand the reasons for selectivity towards other kinases and towards particular BCR-Abl mutants. Structures with novel chemotypes were also used to investigate the conformational flexibility of Abl kinase, and to identify potential new interactions which could be used to increase potency. These structures were followed up by co-crystal structures with modified chemotypes to support medicinal chemistry as part of the structure-based drug design cycle involving computer aided design, chemistry, biochemistry, biology and crystallography. In this article we present the approaches used to obtain multiple co-crystal structures with Abl kinase, the type of information obtained and how this information was used to support drug discovery.

2. Structural biology

Full details of the methods used in this work are presented in Cowan-Jacob et al. (2007), from which this chapter was adapted with permission.

2.1. PROTEIN PREPARATION

The three different constructs of Abl kinase prepared for this study (Fig. 1) show a similar behaviour. Expression in Sf9 insect cells yields around 10 mg/L soluble Abl protein, which is however heterogenously phosphorylated. To obtain homogeneous protein for crystallization the unphosphorylated form of Abl kinase is preferred. The fraction of unphosphorylated protein can be increased through dephosphorylation by protein tyrosine specific phosphatases or more conveniently, by including an Abl kinase inhibitor in the cell culture. In several cases the inhibitor leads in addition to an increase in the expression level. Inhibitor addition is however limited by the toxicity of a particular compound towards the insect cells, e.g. NVP-AFN941 could only be added later during expression and at low concentration, and by the stability of the compound under fermentation conditions. Compounds modified during fermentation (e.g. NVP-AEG082) had to be added during cell lysis. All three constructs are labile in their apo form but can be stabilized by ligands, and the degree of stabilization reflects directly the affinity of a particular ligand towards Abl kinase. Stabilization by a high affinity ligand is particularly crucial in the anion exchange step used to separate residual phosphorylated forms and to achieve high protein concentrations for crystallization. Initially construct A which was available in house, was purified and cleaved with Factor Xa to obtain AAMD-Abl(218-500) which crystallized readily in complex with imatinib. However, these crystals diffracted poorly, possibly due to the elongated N-terminus compared to that of the construct B (Schindler et al., 2000). Limited proteolysis with papain gave an N-terminally truncated protein Abl (227/228- 500) which yielded high resolution structures with imatinib and NVP-AEG082. Based on this result and the N-terminus reported for crystallized mouse-Abl (construct B) construct C was designed. Construct B was crystallized once (with NVP-AFG210) to compare with the in-house constructs A and C, but did not show any advantage. An efficient purification procedure for construct C in the presence of a stabilizing ligand has been established and has allowed the preparation and crystallization of many different complexes besides the one reported here with PD180970.

2.2. CRYSTALLISATION

Throughout this project, numerous iterative refinements were made to the processes. Initially screens were done with hanging drop experiments in 24-well trays and required about 100 μL of protein to screen 96 different buffer conditions. The more recent screens were prepared using a robot to distribute the reservoir solutions (reformatting) into 96-well trays and another robot to prepare sitting drops in these trays, requiring less than 30 μL of protein to test 96 different buffer conditions. Initially, viewing of plates was done manually under a microscope, but more recently the introduction of an automated imaging system allowed the viewing of trays stored at 4°C and 20°C remotely, according to a predefined schedule. The advantages of the robotics can be measured in the larger number of experiments that can be done with the same amount of protein, the reduced manual intervention time, and the simultaneous electronic recording of the experimental details and results.

Crystallisation details for the structures published here are listed in Table 1. Much better results were obtained with crystallization of construct C compared to construct A, and to get diffraction quality crystals with the former, seeding was usually necessary. Among the trends noticed for these and the many other Abl complexes crystallized, was the fact that the ease of crystallization (number and simplicity of optimisation of hits from crystal screens), increased in proportion with the increase in affinity of the inhibitor. The Abl protein was found to be much less stable in the absence of an inhibitor, tending to aggregate and precipitate. Presumably the higher affinity inhibitors stabilized the protein, making it more suitable for crystallisation.

TABLE 1. Crystallisation and data collection statistics.

Crystal complex	Imatinib	NVP-AEG082	NVP-AFN941	NVP-AFG210	PD180970
Construct	A (papain cleaved)	A (papain cleaved)	A (papain cleaved)	B	C
Protein concent'ion	25 mg/mL	30 mg/mL	24 mg/mL	25 mg/mL	28 mg/mL
Crystallisation buffer	16% PEG 8000 0.1 M MES pH 6.75 0.2 M MgAcetate	28% PEG 4000 0.1M Tris.HCl pH 8.0 0.2 M NaAcetate	12% PEG 8000 0.1 M HEPES pH 7.5 0.2 M MgAcetate	16% PEG 8000 0.1 M NaCacodylate pH 6.2 0.2 M MgAcetate	0.9 M NaAcetate 0.1 M NaCacodylate pH 7.0
Cryobuffer	30% glycerol	15% glycerol	25% glycerol	20% glycerol	30% glycerol
Microseeding	Yes	No	Yes	Yes	No
Method	Hanging drop VD	Hanging drop VD	Hanging drop VD	Microbatch under oil	Hanging drop VD
Temperature	4°C	20°C	4°C	4°C	4°C
Synchrotron source	DESY BW7B	ESRF SNBL	SLS PX 06	SLS PX 06	ESRF ID14-1
Temperature	100 K	100 K	100 K	94 K	100 K
Wavelength	0.8452 Å	0.8000 Å	0.9778 Å	0.8602 Å	0.934 Å
Detector	Mar300	Mar345 DTB	MarCCD 165	MarCCD 165	MarCCD 165
Space group	$C222_1$	$P2_12_12_1$	C2	$I4_122$	$P2_12_12$

(Continued)

Table1 (Continued)

Unit cell (Å)	141.7, 148.7, 115.3	34.0, 124.1, 139.5	185.4, 58.9, 104.0 β=119.0	105.4, 105.4, 110.4	106.5, 131.5, 56.5
Contents of asymmetric unit	4	2	3	1	2
VM (solvent content)	2.4 $Å^3$/Da (49%)	2.3 $Å^3$/Da (47%)	2.6 $Å^3$/Da (53%)	2.4 $Å^3$/Da (49%)	3.1 $Å^3$/Da (60%)
Resolution range (high) (Å)	25.0 – 2.4 (2.49–2.40)	25 – 2.1 (2.18–2.10)	40.0 – 2.8 (2.90–2.80)	32.0 – 2.7 (2.80–2.70)	34.3 – 1.7 (1.79–1.70)
Rmerge	0.055 (0.438)	0.087 (0.252)	0.087 (0.335)	0.036 (0.479)	0.070 (0.424)
Completeness (%)	99.5 (100.0)	99.8 (99.9)	96.2 (83.3)	96.0 (60.8)	99.5 (100.0)
Unique reflections	47,314	35,534	22,731	17,626	87,524
Multiplicity	4.1	6.1	2.8	4.0	4.8
Processing program	Denzo/ Scalepack	Denzo/ Scalepack	Denzo/ Scalepack	Denzo/ Scalepack	Mosflm

2.3. STRUCTURE DETERMINATION AND REFINEMENT

The phasing and refinement of the structures presented here was relatively straightforward, except for the NVP-AFG210 and NVP-AFN941 complexes (Table 2). In the case of NVP-AFG210, the overall temperature factor for the data based on Wilson statistics was estimated at 77 Å2, and the individual values refined to an average of 80 Å2 for the protein, with a range of 33–175 Å2. The electron density resembled that of a lower resolution structure, the ligand occupancy was weak, and the R-factors converged at rather high values (Rf = 0.264, R free = 0.314). Water molecules were not added to the electron density due to the low resolution. Repeated data collection from different crystals gave no improvement. The data were checked for twinning and also processed in lower symmetry space groups, but no errors could be identified. There are some close contacts between the inhibitor and the protein and it should be considered a low quality structure, from which only the mode of binding and not detailed interactions can be inferred. In the case of the Abl-NVP-AFN941 complex, there was a technical problem during collection of the data at the synchrotron, in which the total phi range of each image was actually shorter than requested. This data collection was not repeated because other structures had higher priority and the quality of the electron density in this case was reasonable for the resolution of the data obtained. The surprising result for this complex was the fact that only one of the three molecules in the asymmetric unit contained the inhibitor, despite the fact that the crystallization was performed with a stoichiometric ratio of the protein and NVP-AFN941. The other two inhibitors were found to be forming crystal contacts between protein molecules.

TABLE 2. Refinement statistics and inhibition values.

Compound	Imatinib	NVP-AEG082	NVP-AFN941	NVP-AFG210	PD180970
IC_{50} (nM)*					
Abl wt	170 ± 23	330 ± 113	3425 ± 1075 30% at 3 mM	41 (n = 1)	70 ± 32
Bcr-Abl	231 ± 43	565 ± 60		> 7,000	22 ± 2
Model contents:					
Protein atoms	2071, 2098, 2081, 2098	2121, 1966	2139, 2134, 2115	2182	2210, 2117
Inhibitor atoms	37, 37, 37, 37	31, 31	35, 35, 35	27	29, 29
Water molecules	200	387	19	0	540
R-factor	0.208	0.195	0.218	0.264	0.176
Free R-factor	0.260	0.256	0.287	0.314	0.204
R.m.s.d. bonds (Å)	0.02	0.03	0.02	0.01	0.02
R.m.s.d. angles (°)	1.86	1.57	1.87	1.16	1.69
Ramachandran plot (%):					
Most favoured	88.1	87.6	86.3	84.6	88.8
Allowed	11.0	11.1	13.3	13.3	11.2
Generously allowed	0.9	1.1	0.4	2.1	0.0
Disallowed	0.0	0.2**	0.0	0.0	0.0
PDB access codes	2HYY	2HZ0	2HZ4	2HZN	2HZI

*IC50 values represent inhibitor concentrations (nM) required to inhibit activity by 50% (proliferation of Bcr-Abl-dependent murine haematopoietic Ba/F3 cells (see Manley et al., 2005b, for an overview)).
**A234-lys, in weak density at N-terminus.

There is an ongoing debate in the pharmaceutical industry about the balance between quality and impact in structural biology. Considerable efforts can be made to try and improve the crystals to get higher resolution data, to recollect data sets that have lower quality due to a technical problem (e.g. the short phi-range in the NVP-AFN941 complex dataset), and to refine the structure to a point where even the most meticulous crystallographer would not find a better fit. However, what level of quality is really required to be sure that one has an accurate model of the inhibitor binding to the protein target, which can be used as a basis for drug design? Some groups only spend time refining the part of the structure close to the ligand binding site, which in general gives an accurate result when the quality of the data is high and when multiple structures of the same protein have already lead to a very good model. We prefer to refine all structures to convergence, especially in a case like this where many different space groups are found and different crystal environments lead to many different loop conformations, as Abl kinase is a rather flexible protein. But if repeating the experiment requires going back to protein expression in the presence of the inhibitor, or if getting better data means going back to screening to find a new crystal form, and if the quality of the data are sufficient to be sure of the mode of binding of the inhibitor, this extra work may not be justifiable in an environment where time and resources are critical. What is important is to be able to recognise what level of detail the quality of the structure is justified to provide, and not to over interpret the results (Davis et al., 2003).

3. Structure of the imatinib complex and study of mutants

3.1. THE BINDING OF IMATINIB

The structure of human Abl in complex with imatinib (Manley et al., 2002) is very similar to the structure of the mouse enzyme (Nagar et al., 2002), despite the fact that they were obtained in different space groups and with different length proteins (Fig. 1). The average r.m.s.d. value for all 261-264 Cα atoms of pair-wise superimposed subunits (four protein molecules in the human structure, two in the mouse structure) varies between 0.34 and 0.80 Å. The main differences occur in loop regions where the temperature factors are high (e.g. the β3-αC loop and the A-loop), and are not highly correlated with crystal contacts. There are slightly higher differences in the N-terminal lobe, on average, because there is a tendency for this lobe to rotate as a rigid body with respect to the C-terminal lobe.

Imatinib binds in the cleft between the N- and C-terminal lobes, as expected (Fig. 2). However, in contrast to what had been predicted by homology modelling (Zimmermann et al. 2001) but with agreement to that suggested by the crystal structure of Abl in complex with des-methyl-piperazinyl imatinib (Schindler et al., 2000), was that imatinib bound to an inactive conformation where the DFG motif flips out to make a channel beyond the gatekeeper residue (Thr315) to accommodate the benzamide and N-methyl piperazine groups (Nagar et al., 2002; Manley et al., 2002) The rest of the A-loop also adopts an inactive conformation in which the residues around Tyr393 bind in and block the substrate binding site. In addition, the P-loop folds down to form a cage around the pyridine and pyrimidine groups of imatinib. The N-methyl piperazine group, which was incorporated into the molecule during optimization to improve solubility, is found to have a strong interaction with the protein via hydrogen bonds with the main chain carbonyl groups of Ile360 and His361 (Fig. 2). Other hydrogen bond interactions are found between the pyridine-N and the backbone NH of Met318 in the hinge region, the anilino-NH and the side chain of the gatekeeper residue Thr315, the amide-NH and the side chain of Glu286 from helix C and the amide-carbonyl and the backbone-NH of Ala380, which

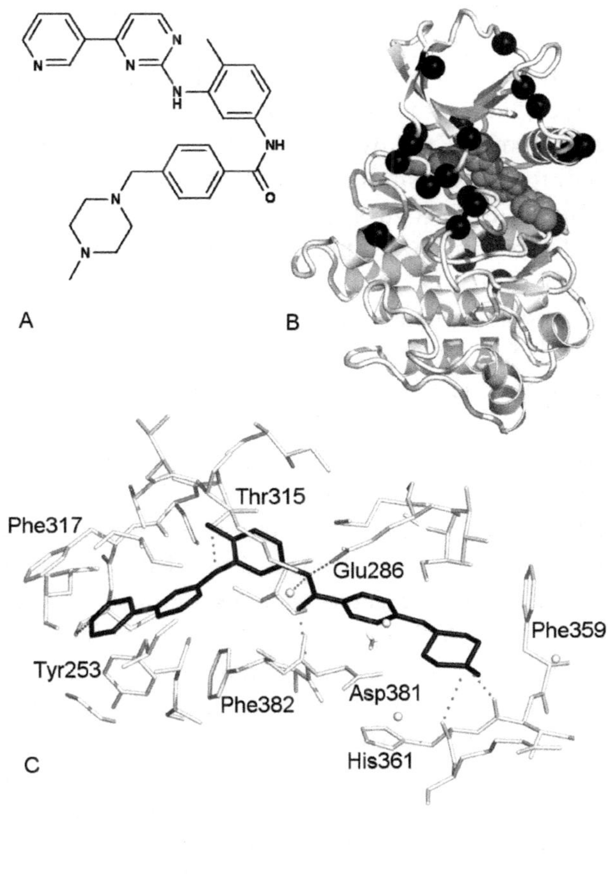

Figure 2. The binding of imatinib to Abl kinase. (a) chemical structure of the inhibitor, (b) overall structure of the kinase domain, with the non-hydrogen atoms of imatinib shown as grey spheres and the locations of mutations arising in relapsed patients as dark spheres, and (c) details of the binding with hydrogen bonds indicated by dashes.

just precedes the highly conserved DFG motif. These hydrogen bonds are complemented by extensive hydrophobic interactions over the whole length of the inhibitor, although less for the N-methyl piperazine which is partially exposed to solvent.

3.2. REASONS FOR SELECTIVITY

Determinants of selectivity in Abl kinase include the conformation of the DFG motif at the start of the A-loop and the conformation of the P-loop. Kinases such as c-Src that are presumably energetically less able to adopt the DFG-out conformation, cannot bind imatinib strongly (Cowan-Jacob et al. 2005). c-Src shares a very high sequence identity with Abl kinase (50% in the kinase domain and about 90% in the ATP site and around the DFG motif), but has a different inactive kinase conformation compared to inactive Abl in complex with imatinib (Xu et al., 1999; Nagar et al., 2002; Nagar et al., 2003). In the assembled inactive c-Src structure, the C-helix is shifted out of the active site and the A-loop forms a single turn of helix that stabilises the position of helix C by packing underneath it. However, the DFG motif has a conformation similar to that observed in active kinases. As a result, imatinib or its analogues bind to c-Src or the homologous Syk with a different and lower affinity binding mode (Cowan-Jacob et al., 2005; Attwell et al., 2004).

The DFG-out conformation has been seen for other kinases such as the Ser/Thr kinases b-Raf (Wan et al., 2004) and P38 Map kinase (Pargellis et al., 2002); as well as for the cKit (Mol et al., 2004), KDR/VEGFR-2 (Manley et al., 2002), Flt-3 (Griffith et al., 2004) and Irk (Hubbard et al., 1994) receptor kinases. However, imatinib does not bind to b-Raf, P38, KDR, Flt-3 or Irk, although it does bind to c-Kit. Therefore, although the ability of the kinase to adopt the DFG-out conformation is necessary for it to bind to imatinib, this is not sufficient. One of the reasons for this is the size of the side-chain in the gatekeeper residue at the back of the ATP-binding site. In Abl the gatekeeper is Thr315, while in Irk it is a methionine, and in Flt-3 it is a phenylalanine. If these side-chains are bigger than threonine, they partially block the path to the pocket formed by the DFG-out conformation, which restricts the types of compounds that can access this pocket. In KDR this residue is a valine, which would not cause steric hindrance, but would be unable to form a hydrogen bond with imatinib. The reason for the lack of binding of imatinib to P38 kinase, which has threonine in the gatekeeper position like Abl, is the existance of a shorter hinge region that changes the shape of the ATP binding site. The fact that imatinib is observed to bind to c-Kit suggests that c-Kit, which also has a threonine in the gatekeeper position and shares high homology to Abl in the

ATP site and the DFG region, can also adopt the DFG-out conformation, and this was recently confirmed by Mol et al. (2004). Presumably PDGFRβ can adopt a similar conformation, although this structure has not yet been published.

The P-loop (phosphate binding, or glycine-rich loop) contributes to the selectivity, because, as mentioned above, in the imatinib complex this loop adopts an inactive conformation that forms extensive contacts with the inhibitor. This conformation allows Tyr253 of the P-loop to form a face-to-edge aromatic interaction with the pyrimidine group of imatinib. A similar conformation has been seen in an FGFR1 kinase–inhibitor complex where it is also thought to contribute to binding of the inhibitor and therefore, selectivity (Mohammadi et al., 1997). In the imatinib-cKit complex the conformation of the P-loop resembles that found in active kinases. This is due to the presence of a cysteine in c-Kit in the place of Ala380 in Abl, which requires a different conformation of the Phe382 side-chain for steric reasons, and allows this side-chain to form a face-to-edge aromatic interaction with imatinib (Mol et al., 2004, SCJ, GF, unpublished data). At the same time, the Phe382 side chain prevents the P-loop from adopting the cage conformation observed in c-Abl. Hence, imatinib will bind strongly to kinases in which the P-loop sequence favours the adoption of a conformation that allows extensive contacts with the inhibitor, or in which other structural elements can form similar interactions. The consequence of these observations for the design of new inhibitors is that incorporating interactions between the inhibitor and the protein that take advantage of special inactive conformations of the kinase is likely to provide selectivity.

3.3. REASONS FOR RESISTANCE

The single site mutations of BCR-Abl were mapped onto the imatinib-Abl complex structure as they were reported (Shah et al., 2002). The surprising observation was that the detected mutations do not cluster around the imatinib binding site, but are spread throughout the kinase domain (Fig. 2). The reasons for resistance have been discussed in depth elsewhere (Cowan-Jacob, 2004), but the general trends are summarised here. The most frequently detected mutations affect the P-loop residues Gly250, Tyr253 and Glu255, Thr315 (the gatekeeper residue in the hinge region), Met351 (distant from the binding site in the C-terminal lobe of the kinase) and Phe359 (from the C-terminal lobe). The Thr315Ile gatekeeper mutation causes steric hindrance with imatinib and the loss of a hydrogen bond, which explains why the affinity of imatinib drops dramatically (IC50 > 10 μM for Thr315Ile, compared to 0.20 μM for wt-Abl). The mutation from Threonine to Isoleucine does not effect the binding of ATP, and there are quite a variety of residues found in this position in other kinases in their wild-type state, suggesting that this will be a hot spot for mutations causing resistance to the binding of many inhibitors of kinases, even in their active state. The Glu255Lys/Val P-loop mutations cause the loss of two hydrogen bonds that stabilize the inactive conformation of this special cage around the inhibitor, and would therefore tend to shift the equilibrium distribution of the kinase conformational states towards the active conformation, to which imatinib does not bind (IC50 value 6.7 μM for Glu255Lys). Other P-loop mutations such as Tyr253His, or Gly250Glu, probably cause imatinib insensitivity (IC50 > 10 μM and 3 μM, respectively) due to either destabilization of the inactive conformation or stabilization of the active conformation of the kinase. For example, the glutamate mutation at the position of Gly250 does not show any reason to destabilise the inactive conformation of the P-loop, but it could form hydrogen bonds that would stabilise the conformation of the P-loop in the activated kinase. The distant mutation, Met351Thr is more difficult to explain. A structure of this mutant in complex with nilotinib (NVP-AMN107), shows that the mutation causes a small rearrangement in the core of the C-terminal lobe of the kinase (Weisberg et al., 2005). This is likely to lead to a higher overall entropy of the structure, which will cost more energy for inhibitor binding. In addition it may lead to a shift in equilibrium between the inactive and active conformations of the kinase. All of these distant mutations, including some in the A-loop such as His396Pro, have a relatively small effect on the binding affinity, with IC50 values only three- to six-fold higher than for wild-type Abl. A recent crystal structure of the latter mutant revealed an active conformation of Abl kinase, despite the lack of phosphorylation of the A-loop tyrosine (Young et al., 2006). This is most likely due to the destabilization of the inactive conformation of the A-loop due to the conformational constraints of the proline, but may be enhanced by the fact that the inhibitor, VX-680, sterically favors the active conformation.

The theory that many of these single site mutations lead to a shift of the Abl kinase conformation towards the active state is supported by the work of Azam and colleagues (2003), in which they used an in vitro screen to find imatinib-resistant mutants in the kinase domain and the N-terminal region of Abl. A mapping of these mutants onto the structure of the assembled inactive state of Abl, which includes the SH3, SH2, linker and kinase domains, shows that many of these mutants will interfere with the domain-domain regulatory interactions that are required to maintain the downregulated state of Abl (Nagar et al., 2003). Other mutations would appear to disfavour a Src-like inactive conformation of Abl that might be necessary for the interconversion between the active and inactive states (Levinson et al., 2006).

The binding of imatinib to the inactive conformation of Abl kinase relies on specific interactions and conformational states that have, in many cases, no constraints in the active conformation. For example, the residues observed to be mutated in the P-loop have no effect on the binding of ATP in the active conformation and result in no significant change to the Km values, but the identity of the residues is important for the binding of imatinib because they are contributing to the stabilization of the inactive conformation (Cowan-Jacob et al., 2004). Therefore, the reasons for selectivity are also a strong contributor to the susceptibility to resistance.

4. Exploring the inhibitor binding site

4.1. STRUCTURES OF ABL IN COMPLEX WITH DIFFERENT CHEMOTYPES

An attractive strategy to overcome or avoid most cases of resistance, would be to administer two drugs in combination, which utilise different binding interactions to inhibit the Abl kinase. In particular, a useful combination could be a compound which binds to the inactive conformation, such as imatinib, with a compound which bound to an active conformation. Examples of chemotypes used as leads for targeting the active conformation are NVP-AFN941 (tetrahydrostaurosporin), and PD180970 (Fig. 3). Structures of these complexes show that both inhibitors bind in the ATP site and form hydrogen bonds with the hinge region, and that they are both within van der Waals distance of the Thr315 gatekeeper residue, although the contacts between PD180970 and Thr315 are much more extensive.

The NVP-AFN941 complex structure essentially resembles that of an active kinase, despite the lack of phosphorylation on the A-loop, although some parts of the A-loop and the P-loop are disordered in the crystals. This structure is very similar to other structures of tyrosine kinases in complex with staurosporin, such as Lck, Zap-70, Syk and Fyn (Zhu et al., 1999; Jin et al., 2005; Atwell et al., 2004; Kinoshita et al., 2006). There are only minor differences in distant loops and the A-loop near the phosphorylation site, because some of these structures are phosphorylated and the Abl/NVP-AFN941 complex is not. The PD180970 structure, which is similar to that of a complex with a related compound published by Nagar et al. (2002), shows an inactive conformation of the P-loop and an unusual conformation of the DFG motif (Fig. 3), but otherwise resembles an active kinase conformation concerning the position of the C-helix and the path of the rest of the A-loop. Tyr393 is located in the same position as the phosphorylated tyrosine in active Lck, and there would clearly be room for a phosphate group bound to Tyr393 and interacting with Arg363 and His396. The conformation of the P-loop resembles that of the complex with imatinib, which shows that this conformation can be adopted with different chemotypes and it is not specifically stabilized by imatinib only. The conformation of the DFG motif involves the flipping over of Asp381 to make a strong hydrogen bond with the main-chain carbonyl of Val299 (Fig. 4). This results in the Asp381 side chain occupying what would be the position of the Phe382 side chain in the active conformation, and the Phe382 side chain flipping over to occupy the site of the Asp381 side chain. This "DFG-flip" conformation puts the Phe382 side chain in van der Waals contact distance of the inhibitor. The buffer used to grow the crystals of the Abl-PD180970 complex crystals has a pH of 7.0, so it is unlikely that it is the crystal buffer that favours protonation of the Asp381 side chain, although it cannot be ruled out that this conformation is not an artefact of the crystallization conditions. A similar conformation of the DFG motif is seen in other Abl structures (Nagar et al., 2002, 2003), and it may represent another natural inactive conformation of the kinase. This "DFG-flip" conformation does not expose the pocket beyond the gatekeeper residue that becomes available for inhibitor binding in the "DFG-out" conformation.

The structure in complex with NVP-AEG082 represents that of a chemical class that binds to the inactive "DFG-out" conformation of Abl kinase, but does not induce or stabilise an inactive conformation of the P-loop (Fig. 5). The trifluoromethylphenyl group makes a very complementary fit to the "DFG-out" pocket and the reverse amide of NVP-AEG082 (compared to the orientation of the amide in imatinib), makes analogous hydrogen bonds to Glu286 and the amide nitrogen of Ala380 to those observed for imatinib. Like for most other kinase inhibitors, NVP-AEG082 forms two hydrogen bonds with the main chain atoms of the hinge region. There are two molecules in the asymmetric unit of the crystals and one of these shows a novel A-loop conformation, while the A-loop is not visible in the other due to disorder. The novel A-loop conformation lies in an intermediate position between the active conformation and the imatinib-bound conformation. The path of this segment departs from the latter conformation at Leu383, superimposes again at Lys400, has Tyr393 exposed at the surface, and shows some weak resemblance to the intermediate conformation observed in partially phosphorylated Igf1r kinase (Pautsch et al., 2001). However, this conformation is stabilized by crystal contacts in the Abl structure, so it is not clear if this is really a natural inactive state of the A-loop of Abl kinase. Another difference between the conformation of the protein in the NVP-AEG082 complex and the imatinib complex is the relative position of the N- and C-terminal lobes of the kinase. There is a shift of main-chain atoms in the N-terminal lobe as great as 2.6 Å between the two structures. This seems to be induced by the differing shapes of the inhibitors rather than crystal packing forces, because both molecules in the asymmetric unit of the

NVP-AEG082 complex have different crystal packing contacts, yet they both have the same relative orientation of the N- and C-terminal lobes.

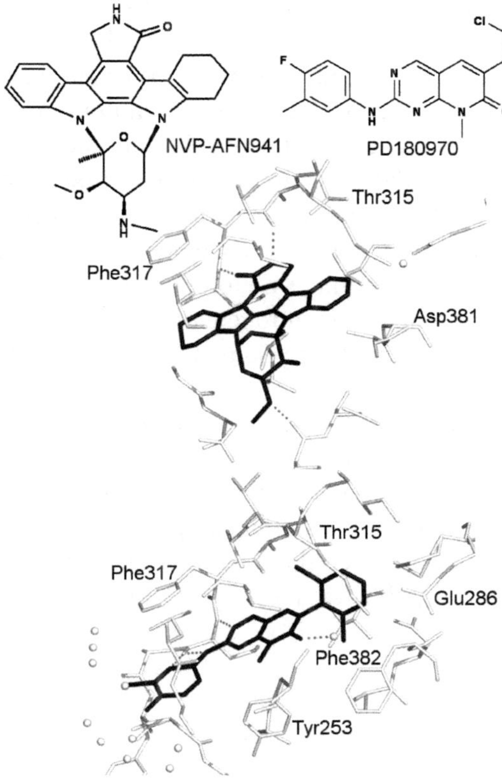

Figure 3. The binding of NVP-AFN941 and PD180970 to Abl kinase. The chemical structures are shown at the top and details of the binding with hydrogen bonds indicated by dashed lines, are shown below.

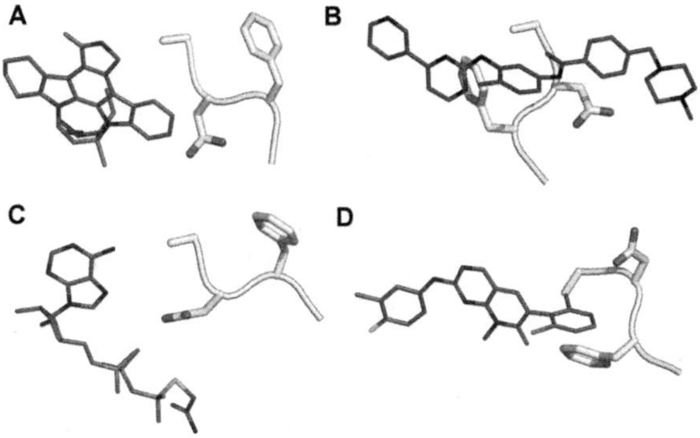

Figure 4. Conformations of the DFG-motif, which is located at the N-terminus of the A-loop, observed in various Abl kinase structures to date. (a) active, (b) DFG-out, (c) c-Src-like inactive (PDB entry 1G1T), (d) DFG-flip.

NVP-AFG210, a Raf kinase inhibitor (Thaimattam et al., 2004), which is also active against KDR and Abl, is another example of a compound that binds to the DFG-out conformation of Abl kinase. This chemotype is interesting because it would be predicted to bind to the Thr315Ile gatekeeper mutant, and although this particular compound was not tested against the mutant, other compounds in this chemical series show inhibition of Thr315Ile Abl in the nanomolar range. The central phenyl ring lies more than 1.2 Å further away from Thr315 than the other inhibitors reported here, which leaves enough space for an extra methyl group (Fig. 6). The chemical structure of NVP-AFG210 allows this relative displacement while retaining hydrogen bonding interactions with the hinge region. The pyridine group forms one hydrogen bond to the backbone nitrogen of Met318 and there is also a favourable interaction between the pryridine CH and the main-chain carbonyl of Glu316. The trifluoromethyl benzene binding in the DFG-out pocket superimposes very well with the same group in other inhibitors, despite the difference in the central part of the inhibitor. The urea group forms hydrogen bonds with Glu286 and the main chain of Asp381, although only one of the nitrogens in the urea is participating. The medicinal chemistry for a related series of urea based compounds, which produced potent inhibitors of Abl and PDGFR, was published recently (Manley et al., 2004a).

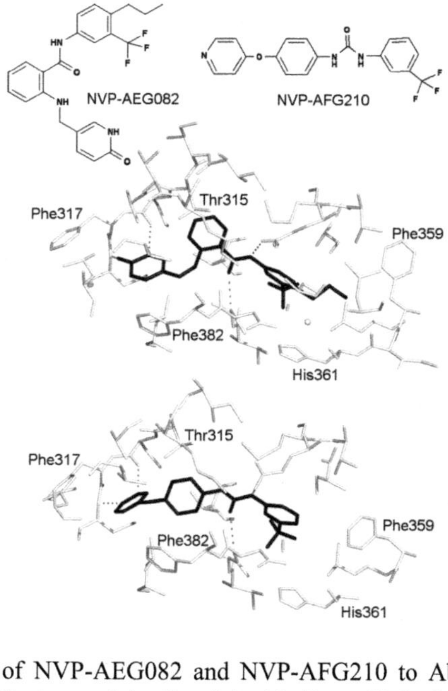

Figure 5. The binding of NVP-AEG082 and NVP-AFG210 to Abl kinase. The chemical structures are shown at the top and details of the binding with hydrogen bonds indicated by dashed lines, are shown below.

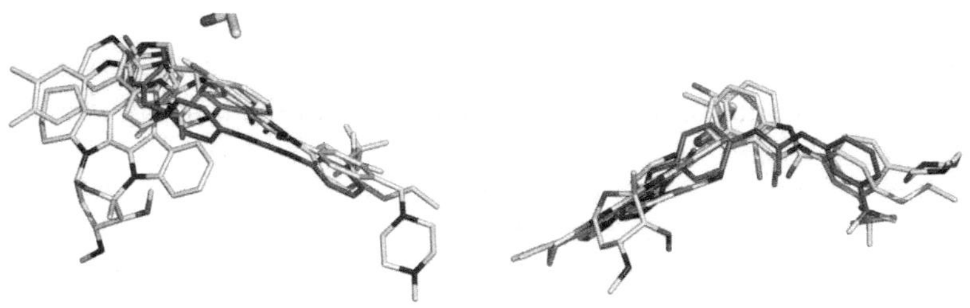

Figure 6. Two views 90° apart around a horizontal axis of the superposition of the five kinase inhibitors mentioned in the text. The structure of NVP-AFG210 is shown in dark grey and the others are light grey. The side chain of Thr315 is also shown, but only visible in the view on the left.

4.2. CONFORMATIONAL FLEXIBILITY OF THE ABL KINASE DOMAIN

Superposition of the various structures reported here, and one of the structures recently reported by Levinson et al. (2006, PDB entry 2G1T), shows that there are four main regions of conformational flexibility in the Abl kinase domain: the A-loop, the P-loop, the C-helix and the relative position of the N-terminal lobe with respect to the C-terminal lobe. These conformational differences cause changes in the properties of the inhibitor binding site and can be exploited to gain selectivity while optimising affinity. Concerning the A-loop, the conformation of the DFG motif has the greatest effect on the binding pocket. There are now four main conformations observed for this highly conserved structural element: the active conformation, the DFG-out conformation, the DFG-flip conformation and the Src-like inactive conformation (Fig. 4, Table 3). A recent publication used molecular dynamics simulations to show that the Src-like inactive conformation of Abl kinase might be an intermediate step for the transition between the active and the inactive conformations of the DFG motif, however, the DFG-flip conformation was not mentioned in this work (Levinson et al., 2006). When comparing the surfaces of the various structures, it can be seen that there are small pockets under the C-helix in the structure with the DFG motif in the active conformation (NVP-AFN941 complex) and these are larger in the structure with the flipped conformation of the DFG motif (PD180970), suggesting that this could also be an intermediate state between the active and DFG-out conformations (Fig. 4). The backbone Phi and Psi angles of the residues defining the DFG motif conformations show that the major difference between the DFG-flip conformation and the others is a rotation of about 70° about the Ala380 Phi angle, while the Asp381 angles are somewhere between the DFG-out and the Src-like inactive conformations and the Phe382 angles are much the same as the Src-like inactive conformation. Concerning inhibitor binding, the DFG-flip conformation allows Phe382 to make interactions with the inhibitor, and changes the shape of the binding site.

TABLE 3. Backbone torsion angles that define the conformation of the DFG motif.

DFG	Ala380		Asp381		Phe382	
conformation	Phi	Psi	Phi	Psi	Phi	Psi
Active (n = 3)	–130	180	60	80	–90	10
DFG-out (n = 7)	–140	–170	–150	110	–90	–10
DFG-flip (n = 2)	–70	140	–120	20	–70	150
2G1T (n = 2)	–130	170	50	40	–70	130
2G2H-A (n = 1)	–70	140	–120	20	–80	–30
2G2H-B (n = 1)	–60	140	–80	–20	–80	170

The conformation of the P-loop has an affect on the properties of the inhibitor binding site at the entrance of the ATP pocket (left hand side of the binding site in the figures). In the cases where the P-loop forms a cage around the inhibitors, the entrance is almost closed, while in the NVP-AFN941 complex it adopts an active conformation and is partially disordered, so the entrance is quite open. The NVP-AEG082 complex has the extended conformation of the P-loop typical of the active conformation, but the entrance to the ATP site looks quite closed compared to the other complexes. This is because Phe382 of the DFG motif adopts a conformation similar to that observed in the c-Kit–imatinib complex, where it extends across underneath the P-loop toward the entrance of the ATP pocket and forms a hydrophobic surface that is complementary to the shape of the inhibitor. NVP-AEG082 does not extend far toward the entrance of the ATP pocket, so does not require space in this region like NVP-AFN941. These observations tend to suggest that the shape of the pocket created by the P-loop is correlated to the conformation of the DFG motif, which is because they contact each other in the structures.

All of the compounds reported here occupy different parts of the binding site. For example, the dichlorophenyl group of PD180970 has a very good fit to the pocket lined by the gatekeeper residue, which is not used by NVP-AFN941. Imatinib fills this region with a methylbenzene group in a very similar way to PD180970, while AEG082 also fits it well, due to the ortho-substitution pattern of the benzene ring. NVP-AEG082 and NVP-AFG210 have an excellent fit to the DFG-out pocket, while imatinib does not have good shape complementarity in this region and NVPAFN941 and PD180970 don't use this pocket at all. A comparison of the way different chemotypes use the binding site allows one to design new structures that use all the subsites in the most effective way, which can be used to gain affinity and selectivity (see Liu and Gray 2006, for an example).

4.3. STRUCTURES TO SUPPORT CHEMICAL OPTIMISATION

Based on the structures of imatinib and other lead series bound to Abl kinase, many suggestions were made for synthesis of new and improved compounds with the aim of gaining potency and at least retaining selectivity. Only one of these paths will be presented here briefly: Four different groups were proposed to take advantage of the hydrogen bonds observed for the amide group of inhibitors to Glu286 and Ala381. The formation of an amide, a reverse amide, a sulfonamide and a urea, were the basis of the creation of chemical libraries to explore binding in the DFG-out pocket, while retaining the methylphenyl-(pyridinyl-pyrimidinyl)-amine moiety of imatinib in the ATP pocket (Manley et al., 2004a, b).

Compounds giving IC50 values against c-Abl in the nM range were found for each series except for the sulfonamides, where all inhibitors were inactive. Structures were determined for certain of these potent inhibitors and used as a basis for ideas to support the optimization of chemical properties such as chemical stability, metabolic stability, and solubility while retaining affinity (PWM, et al., to be published elsewhere; Manley et al., 2004a, b). Incorporation of a hydroxyl group into the reverse amides series to target Asp381 gave a compound (NVP-AHT202) having an excellent kinase inhibition profile, but poor drugability characteristics (PWM, et al., Structure to be presented elsewhere). The investigation of alternative donor/acceptor groups finally led to the synthesis of NVP-AMN107 (nilotinib), which is highly potent, very selective, active against all but one of the resistance mutants isolated from relapsed imatinib patients, has good pharmacokinetic properties and is now in clinical trials in man (Weisberg et al., 2006; Kantarjian et al., 2006). Structures of nilotinib in complex with Abl show that it makes similar hydrogen bond interactions to imatinib with Abl kinase, except for those formed by the N-methyl piperazine group of imatinib to the C-terminal lobe (Fig. 2; Weisberg et al., 2005). Nilotinib has a better fit to the DFG out pocket than imatinib, which probably accounts largely for the increased affinity (Fig. 7). There is also a weak electrostatic interaction between a fluorine from the trifluoro-methyl group of nilotinib and the polarized carbon of the Ala380 carbonyl group (Manley et al., 2005b).

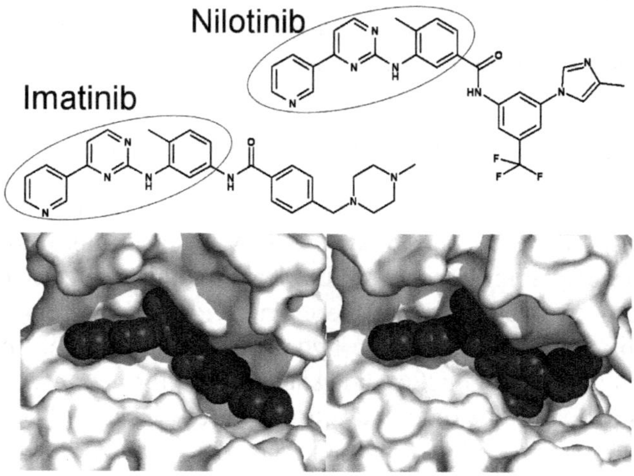

Figure 7. Top: chemical structures of imatinib and nilotinib with the common moieties circled. *Bottom*: comparison of imatinib (*left*) and nilotinib (*right*) binding to Abl kinase. The parts of the surface contributed by the P-loop and the DFG motif have been left out for clarity.

5. Nilotinib and it's ability to overcome most imitinib-resistant mutant forms of BCR-Abl

The increased potency of nilotinib as a selective Abl/BCR-Abl kinase inhibitor seems to result from the improved lipophilic interactions, and a reduced need for energetically expensive desolvation compared to the N-methyl piperazine group of imatinib, which is highly basic and therefore protonated at physiological pH. As a consequence of the enhanced binding to the Abl kinase domain, the increased potency leads to greater reduction in tumor burden in patients (less BCR-Abl expressing cells), and therefore, there are less cells available to develop resistance (O'Hare et al., 2007). In addition, the increased affinity leads to the ability to inhibit most of the imatinib resistant mutants, allowing it to overcome the slight shift of the equilibrium from the inactive toward the active caused by the mutations. Thus in *in vitro* studies, nilotinib has been shown to inhibit 32 of 33 clinically relevant imatinib-resistant mutations at physiologically relevant concentrations, with the IC_{50} values for BCR-Abl-dependent cell proliferation being below 1 μM compared to trough concentrations of nilotinib in the blood of > 1.5 μM following dosing at 400 mg twice-daily. In fact, 29 of these mutations are sensitive to nilotinib with IC_{50} values of less than 300 nM. The three least sensitive mutations are located in the P-loop, which has a similar inactive conformation as to when imatinib is bound. The remaining mutation, T315I, is not sensitive to nilotinib, or any of the other Abl inhibitors currently marketed for the treatment of CML, because of a steric clash between the side chain of the substituted residue and the inhibitor.

As predicted from the preclinical profile, nilotinib has shown good efficacy in imatinib-intolerant and imatinib-resistant patients with either chronic- or advanced phase CML (Kantarjian et al., 2007; le Coutre et al., 2008). When administered at the recommended dose of 400 mg twice-daily, nilotinib was well tolerated in these patient populations, with adverse events being mostly mild to moderate and generally reversible and manageable with symptomatic treatment. Nilotinib (Tasigna®) was first approved in Switzerland (July 2007) for use in treatment of patients and has subsequently been approved in many more countries including the United States and European Union.

6. Conclusions

As stated by Levinson et al. (2006), working with an unphosphorylated protein seems to allow the sampling of numerous possible conformational states. Phosphorylation will stabilise an active conformation, but the binding of an inhibitor or even the presence of different types of crystal environments can stabilise the active and other conformations of the unphosphorylated protein. Working with unphosphorylated protein does not preclude seeing the active conformation. It is even possible to isolate a protein in two different conformational states, or different ligation states, within the same crystal (e.g. the NVP-AFN941 complex reported here, and Levinson et al. 2006). An examination of the different conformational states of Abl kinase has lead to an excellent understanding of the features of the binding important for potency, selectivity and susceptibility to resistance.

After imatinib became the first successful small molecule kinase inhibitor to reach the market (Capdeville et al., 2002), publication of the first structure of Abl kinase (Schindler et al., 2000), and recognition that mutant Bcr-Abl was a key mechanism of imatinib-resistance (Shah et al., 2002), the race was on to use structural biology to help find even better treatments for CML. As a result, the structural biology program experienced considerable pressure to get novel structures quickly. This pressure is further emphasized by the large number of publications which illustrate the competitive nature of the field. In order to get the structural results rapidly, the use of protein produced from multiple constructs, the use of crystallisation robotics, and the use of partially automated software was important. The availability of structures, and especially the fact that these structures showed a mode of binding that was not predicted by homology modeling at the time, considerably improved the understanding of how the drug worked. Follow-up structures of different Abl kinase conformations induced by inhibitors with a variety of chemotypes, provided a basis for de novo design, data-base mining and virtual screening to find new chemical structures which could bind to Abl. X-ray analysis of crystalline complexes of these leads bound to the Abl kinase domain then provided details of the binding modes of these new chemotypes, which allowed further optimization of their structures for potency against Bcr-Abl and it's mutants, selectivity over other kinases and general drugability properties. Structural biology was a strong contributor to the quick discovery of nilotinib (NVP-AMN107), which was first synthesized in 2002 and entered clinical trials in 2005 (Kantarjian et al., 2006).

ACKNOWLEDGEMENTS

The authors would like to acknowledge the support of the staff of the following synchrotron beamlines: DESY BW7B (Hamburg, DE), ESRF SNBL, ESRF ID14-1 (Grenoble, FR) and SLS PX06 (Villigen, CH). All pictures have been prepared using the program Pymol (http:\\ pymol.sourceforge.net\).

References

Attwell, S., Adams, J.M., Badger, J., Buchanan, M.D., Feil, I.K., Froning, K.J., Gao, X., Hendle, J., Keegan, K., Leon, B.C., Muller-Deickmann, H.J., Nienaber, V.L., Noland, B.W., Post, K., Rajashankar, K.R., Ramos, A., Russell, M., Burley, S.K. & Buchanan, S.G., 2004, A novel mode of Gleevec binding is revealed by the structure of spleen tyrosine kinase, *J. Biol. Chem.* 279:55827–55832.

Azam, M., Latek, R.R. & Daley, G.Q., 2003, Mechanisms of autoinhibition and STI-571/imatinib resistance revealed by mutagenesis of BCR-ABL. Cell 112:831–843.

Baccarani, M., Saglio, G., Goldman, J., Hochhaus, A., Simonsson, B., Appelbaum, F., Apperley, J., Cervantes, F., Cortes, J., Deininger, M., Gratwohl, A., Guilhot, F., Horowitz, M., Hughes, T., Kantarjian, H., Larson, R., Niederwieser, D., Silver, R. & Hehlmann, R., 2006, Evolving concepts in the management of chronic myeloid leukemia: recommendations from an expert panel on behalf of the European LeukemiaNet, *Blood*, 108:1809–1820.

Capdeville, R., Buchdunger, E., Zimmermann, J. & Matter, A., 2002, Glivec (STI571, imatinib), a rationally developed, targeted anticancer drug, *Nat. Rev. Drug Discov.* 1:493–502.

Cowan-Jacob, S.W., Guez, V., Fendrich, G., Griffin, J.D., Fabbro, D., Furet, P., Liebetanz, J., Mestan, J. & Manley, P.W., 2004, Imatinib (STI571) resistance in chronic myelogenous leukemia: molecular basis of the underlying mechanisms and potential strategies for treatment. *Mini-Rev. Med. Chem.* 4:285–299.

Cowan-Jacob, S.W., Fendrich, G., Manley, P.W., Jahnke, W., Fabbro, D., Liebetanz, J. & Meyer, T., 2005, The crystal structure of a c-Src complex in an active conformation suggests possible steps in c-Src activation, *Structure* 13:861–871.

Cowan-Jacob, S.W., Fendrich, G., Floersheimer, A., Furet, P., Liebetanz, J., Rummel, G., Rheinberger, P., Centeleghe, M., Fabbro, D., Manley, P.W., 2007, Structural biology contributions to the discovery of drugs to treat chronic myelogenous leukemia. *Acta Cryst.*, D63:80–93.

Davis, A.M., Teague, S.J. & Kleywegt, G.J., 2003, Application and limitations of X-ray crystallographic data in structure-based ligand and drug design, *Angew. Chem. Int. Ed. Engl.* 42:2718–2736.

Druker, B.J., 2006, Cirumventing resistance to kinase-inhibitor therapy, *New Engl. J. Med.* 354:2594–2596.

Gorre, M.E., Mohammed, M., Ellwood, K., Hsu, N., Paquette, R., Rao, P.N. & Sawyers, C.L., 2001, Ckinical resistance to STI-571 cancer therapy caused by BCR-ABL gene mutation or amplification, *Science* 293:876–880.

Griffith, J., Black, J., Faerman, C., Swenson, L., Wynn, M., Lu, F., Lippke, J. & Saxena, K., 2004, Structure of FLT3 autoinhibited by the juxtamembrane domain: implications for acute myeloid leukemia, *Mol. Cell* 13:169–178.

Hubbard, S.R., Wei, L., Ellis, L., & Hendrickson, W.A., 1994, Crystal structure of the tyrosine kinase domain of the human insulin receptor, *Nature* 372:746–753.

Jin, L., Pluskey, S., Petrella, E.C., Cantin, S.M., Gorga, J.C., Rynkiewicz, M.J., Pandey, P., Strickler, J.E., Babine, R.E., Weaver, D.T. & Seidl, K.J., 2005, The three-dimensional strucutre of the Zap-70 kinase domain in complex with staurosporine. Implications for the design of selective inhibitors, *J. Biol. Chem.* 279:42818–42825.

Kantarjian, H.M., Giles, F., Gattermann, N., Bhalla, K., Alimena, G., Palandri, F., Ossenkoppele, G.J., Nicolini, F.-E., O'Brien, S.G., Litzow, M., Bhatia, R., Cervantes, C., Haque, A., Shou, Y., Resta, D.J., Weitzman, A., Hochhaus, A. & le Coutre, P., 2007, Nilotinib (formerly AMN107), a highly selective BCR-ABL tyrosine kinase inhibitor, is effective in patients with Philadelphia chromosome–positive chronic myelogenous leukemia in chronic phase following imatinib resistance and intolerance, *Blood* 110:3540–3546.

Kantarjian, H. M., Giles, F., Wunderle, L., Bhalla, K., O'Brien, S., Wassman, B., Tanaka, C., Manley, P., Rae, P., Mietlowski, W., Bochinski, K., Hochhaus, A., Griffin, J.D., Hoelzer, D., Albitar, M., Dugan, M., Cortes, J., Alland, L. & Ottmann, O.G., 2006, Nilotinib in imatinib-resistant CML and Philadelphia chromosome-positive ALL, *N. Engl. J. Med.* 354:2542–2551.

Kinoshita, T., Matsubara, M., Ishiguro, H., Okita, K. & Tada T., 2006, Structure of human Fyn kinase domain complexed with staurosporine, *Biochem. Biophys. Res. Comm.* 346:840–844.

le Coutre, P., Ottmann, O.G., Giles, F., Kim, D.-W., Cortes, J., Gattermann, N., Apperley, J.F., Larson, R.A., Abruzzese, E., O'Brien, S.G., Kuliczkowski, K., Hochhaus, A., Mahon, F.-X., Saglio, G., Gobbi, M., Kwong, Y.-L., Baccarani, M., Hughes, T., Martinelli, G., Radich, J.P., Zheng, M., Shou, S., & Kantarjian, H., 2008, Nilotinib (formerly AMN107), a highly selective BCR-ABL tyrosine kinase inhibitor, is active in patients with imatinib-resistant or –intolerant accelerated-phase chronic myelogenous leukemia, *Blood* 111:1834–1839.

Levinson, N.M., Kuchment, O., Shen, K., Young, M.A., Koldobskiy, M., Karplus, M., Cole, P.A. & Kuriyan J., 2006, A Src-like inactive conformation in the Abl tyrosine kinase domain, *PLoS Biol.* 4:753–767.

Liu, Y. & Gray, N.S., 2006, Rational design of inhibitors that bind to inactive kinase conformations, *Nat. Chem. Biol.* 2:358–364.

Manley, P.W., Cowan-Jacob, S.W., Buchdunger, E., Fabbro, D., Fendrich, G., Furet, P., Meyer, T. & Zimmermann, J., 2002, Imatinib: A selective tyrosine kinase inhibitor, *Eur. J. Cancer* 38(Suppl 5): S19–S27.

Manley, P.W., Bold, G., Brüggen, J., Fendrich, G., Furet, P., Mestan, J., Schnell, C., Stolz, B., Meyer, T., Meyhack, B., Stark, W., Strauss, A. & Wood J., 2004a, Advances in the structural biology, design and clinical development of VEGF-R kinase inhibitors for the treatment of angiogenesis, *Biochim. Biophys. Acta* 1697:17–27.

Manley, P.W., Breitenstein, W., Brüggen, J., Cowan-Jacob, S.W., Furet, P., Mestan, J. & Meyer, T., 2004b, Urea derviatives of STI571 as inhibitors of Bcr-Abl and PDGFR kinases, *Biorg.Med.Chem.Lett.* 14:5793–5797.

Manley, P.W., Cowan-Jacob, S.W. & Mestan, J., 2005a, Advances in the structural biology, design and clinical development of Bcr-Abl inhibitors for the treatment of chronic myelogenous leukemia, *Biochim. Biophys. Acta* 1754:3–13.

Manley, P.W., Cowan-Jacob, S.W., Fendrich, G. & Mestan, J., 2005b, Molecular Interactions between the Highly Selective pan-Bcr-Abl Inhibitor, AMN107, and the Tyrosine Kinase Domain of Abl., *Blood* (ASH Annual Meeting Abstracts), 106: 3365.

Melo, J.V. & Barnes, D.J., 2007, Chronic myeloid leukaemia as a model of disease evolution in human cancer, *Nat. Rev. Cancer* 7:441–453.

Mohammadi, M., McMahon, G., Sun, L., Tang, C., Hirth, P., Yeh, B. K., Hubbard, S.R. & Schlessinger, J., 1997, Structure of the tyrosine kinase domain of fibroblast growth factor receptor in complex with inhibitors, *Science* 276:955–960.

Mol, C.D., Dougan, D.R., Schneider, T.R., Skene, R.J., Kraus, M.L., Scheibe, D.N., Snell, G.P., Zou, H., Sang, B.-C. & Wilson K.P., 2004, Structural basis for the autoinhibition and STI-571 inhibition of cKit tyrosine kinase, *J. Biol. Chem.* 279:31655–31663.

Nagar, B., Bornmann, W.G., Pellicena, P., Schindler, T., Veach, D.R., Miller, W.T., Clarkson, B. & Kuriyan J., 2002, Crystal structures of the kinase domain of c-Abl in complex with the small molecule inhibitors PD173955 and imatinib (STI-571), Cancer Res. 62:4236–4243.

Nagar, B., Hantschel, O., Young, M.A., Scheffzek, K., Veach, D., Bornmann, W., Clarkson, B., Superti-Furga, G. & Kuriyan, J., 2003, Structural basis for the autoinhibition of c-Abl tyrosine kinase, *Cell* 112:859–871.

Nagar, B., Hantschel, O., Seeliger, M., Davies, J.M., Weis, W.I., Superti-Furga, G. & Kuriyan, J., 2006, Organization of the SH3-SH2 unit in active and inactive forms of the c-Abl tyrosine kinase, *Mol. Cell* 21:787–798.

O'Hare, T., Eide, C.A. & Deininger, M.W.N., 2007, Bcr-Abl kinase domain mutations, drug resistance, and the road to a cure for chronic myeloid leukemia, *Blood* 107:2242–2249.

Pargellis, C., Tong, L., Churchill, L., Cirillo, P.F., Gilmore, T., Graham, A.G., Grob, P.M., Hickey, E.R., Moss, N., Pav, S. & Regan, J., 2002, Inhibition of p38 MAP kinase by utilizing a novel allosteric binding site, *Nat. Struct. biol.*, 9:268–272.

Pautsch, A., Zoephel, A., Ahorn, H., Spevac, W., Hauptmann, R. & Nar H., 2001, crystal structure of the bisphosphrylated IGF-1 receptor kinase: insight into domain movements upon kinase activation, *Structure* 9:955–965.

Ren, R., 2005, Mechanism of Bcr-Abl in the pathogenesis of chronic myelogenous leukemia, *Nat Rev Cancer* 5:172–183.

Schindler, T., Bornmann, W., Pellicena, P., Miller, W. T., Clarkson, B. & Kuriyan, J., 2000, Structural mechanism for STI-571 inhibition of abelson tyrosine kinase, *Science* 289:1938–1942.

Shah, N.P., Nicoll, J.M., Nagar, B., Gorre, M.E., Paquette, R.L., Kuriyan, J. & Sawyers, C.L., 2002, Multiple BCR-ABL kinase domain mutations confer polyclonal resistance to the tyrosine kinase inhibitor imatinib (STI571) in chronic phase and blast crisis chronic myeloid leukemia. *Cancer Cell* 2:117–125.

Thaimattam, R., Daga, P., Abdul Rajjak, S., Banerjee, R. & Iqbal, J., 2004, 3D-QSAR CoMFA, CoMSIA studies on substituted ureas as Raf-1 kinase inhibitors and its confirmation with structure-based studies, *Bioorg. Med. Chem.* 12:6415–6425.

Vagin, A. & Teplyakov, A., 1997, MOLREP: an automated program for molecular replacement, *J. Appl. Cryst.* 30:1022–1025.

Vajpai, N., Strauss, A., Fendrich, G., Cowan-Jacob, S.W., Manley, P.W., Grzesiek, S., Jahnke, W., 2008, Solution conformations and dynamics of ABL kinase inhibitor complexes determined by NMR substantiate the different binding modes of imatinib/nilotinib and dasatinib, *J. Biol. Chem.* 283:18292–18302.

Wan, P.T.C., Garnett, M.J., Roe, S.M., Lee, S., Niculescu-Duvaz, D., Good, V.M., Cancer Genome Project, Jones, C.M., Marchall, C.J., Springer, C.J., Barford, D. & Marais, R., 2004, Mechanism of activation of the RAF-ERK signalling pathway by oncogenic mutations of B-RAF, *Cell* 116:855–867.

Weisberg, E.L., Manley, P.W., Breitenstein, W., Brueggen, J., Cowan-Jacob, S. W., Ray, A., Huntly, B., Fabbro, D., Fendrich, G., Hall-Meyers, E., Kung, A.L., Mestan, J., Daley, G.Q., Callahan, L., Catley, L., Cavazza, C., Azam, M., Neuberg, D., Wright, R.D., Gilliland, G.D. & Griffin, J.D., 2005, Charaterization of AMN107, a selective inhibitor of native and mutant Bcr-Abl, *Cancer Cell* 7:129–141.

Weisberg, E., Manley, P., Mestan, J., Cowan-Jacob, S., Ray, A. & Griffin, J.D., 2006, AMN107 (nilotinib): a novel and selective inhibitor of BCR-ABL, *Br. J. Cancer* 94:1765–1769.

Xu, W., Doshi, A., Lei, M., Eck, M. J. & Harrison, S. C., 1999, Crystal structures of c-Src reveal features of its autoinhibitory mechanism, *Mol. Cell* 3:629–638.

Young, M.A., Shah, N.P., Chao, L.H., Seeliger, M., Milanov, Z.V., Biggs III, W.H., Treiber, D.K., Patel, H.K., Zarrinkar, P.P., Lockhart, D.J., Sawyers, C.L. and Kuriyan, J., 2006, Structure of the kinase domain of an imatinib-resistant Abl mutant in complex with the Aurora kinase inhibitor VX-680, *Cancer Res.* 66:1007–1014.

Zhu, X., Kim, J.L., Rose, P.E., Stover, D.R., Toledo, L.M., Zhao, H. & Morgenstern, K.A., 1999, Structural analysis of the lymphocyte-specific kinase Lck in complex with non-selective and Src family selective kinase inhibitors, *Structure* 7:651–661.

Zimmermann, J., Furet, P., Buchdunger, E., 2001, STI571 - A new treatment modality for CML? In: Ojima I, Vite G, Altmann K (Eds.), Anticancer Agents: Frontiers in Cancer Chemotherapy. ACS Symposium Series 796. Washington, DC: *American Chemical Society*, 245–259.

INTEGRATING CRYSTALLOGRAPHY INTO EARLY METABOLISM STUDIES

GABRIELE CRUCIANI*, YASMIN ARISTEI,
LAURA GORACCI, EMANUELE CAROSATI
Laboratory for Chemometrics and Cheminformatics, Chemistry Department, University of Perugia, Via Elce di sotto 10, Perugia, Italy

Abstract. Since bioavailability, activity, toxicity, distribution, and final elimination all depend on metabolic biotransformations, it would be extremely advantageous if this information to be produced early in the discovery phase. Once obtained, researchers can judge whether or not a potential candidate should be eliminated from the pipeline, or modified to improve chemical stability or safety. The use of in silico methods to predict the site of metabolism in Phase I cytochrome-mediated reactions is a starting point in any metabolic pathway prediction. This paper presents a new method, which provides the site of metabolism for any CYP-mediated reaction acting on unknown substrates. The methodology can be applied automatically to all the cytochromes whose Xray 3D structure is known, but can be also applied to homology model 3D structures. The fully automated procedure can be used to detect positions that should be protected in order to avoid metabolic degradation, or to check the suitability of a new scaffold or pro-drug. Therefore the procedure is also a valuable new tool in early ADME-Tox, where drug-safety and metabolic profile patterns must be evaluated as soon, and as early, as possible.

Keywords: Metabolic hotspots, Site of metabolism prediction, CYP isoform specificity, Metabolic stability, CYP inhibition, Mechanism based inhibition, Chemical reactivity, CYP reactivity, Rate of metabolism, Prediction of metabolic pathway, CYP Xray structures

* To whom correspondence should be addressed. Gabriele Cruciani, Laboratory for Chemometrics and Cheminformatics, Chemistry Department, University of Perugia, via Elce di Sotto, 10, 06123-Perugia, Italy; e-mail: gabri@chemiome.chm.unipg.it

J.L. Sussman and P. Spadon (eds.), *From Molecules to Medicines*,

1. Introduction

The experimental elucidation of the site of metabolism (i.e. the place in a molecule where the metabolic reaction occurs) is usually a high resource-demanding task, which requires an identifiable isotope in the drug,[1] several experimental techniques,[2] and consumes a considerable amount of compound. Nevertheless, the recognition of the site of metabolism could be a significant advance in designing new compounds with better pharmacokinetic profile.[3]

Labile compounds can be stabilized when the place of metabolism is known by adding stable groups at metabolically susceptible position. It is sometimes possible to remove, replace, or protect metabolically susceptible groups. Toxic metabolites in drug candidates can be avoided by chemically protecting the labile moieties. In scaffold hopping and scaffold optimization the recognition of the site of metabolism is crucial in avoiding chemical modification inducing substrate selectivity toward some human cytochromes.

Knowledge of where functional groups are metabolized can help in to design more stable drugs. Figure 1a reports a very potent and selective h5-HT_{2A} receptor antagonist developed by Rowley et al.[4] The bioavailability of fluoropiperidin-phenylindole was 18% and the terminal half-life 1.4 h. The poor pharmacokinetic behaviour was examined carefully, and a major metabolite, the 6-hydroxyindole, was isolated. By blocking the major site of metabolism for the compound in Fig. 1a, using the 6-fluoro derivative reported on Fig. 1b, the pharmacokinetics were dramatically improved when bioavailability increased to 80% and half-life to 12.0 h.

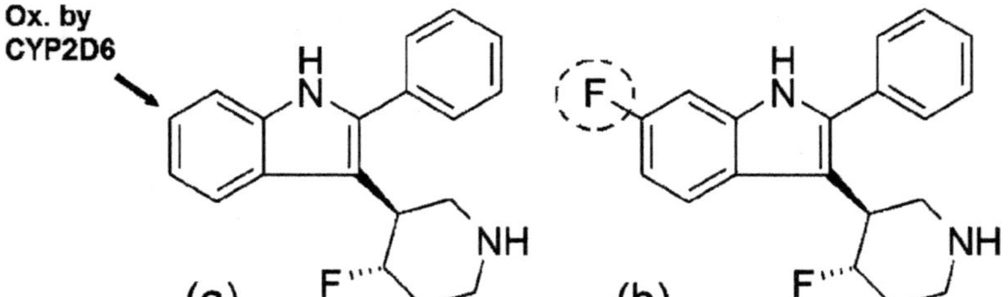

Figure 1. The 3-(4-F piperidin-3-yl)-2-phenyl-1Hindole (a) and the 6-fluoro derivative (b) used by Rowley et al.[4] to block the 2D6 hydroxylation.

To this end the work in research laboratories and drug industries may be largely facilitated by computational predictive methods able to identify the potential site of metabolism of given drug candidates as early in the drug discovery process as possible.

The aim of the present paper is to describe a method[5–7] which is fast, easy to use, computationally inexpensive, and able to predict CYP regio-selective metabolism using only the 3D structures of the CYP enzymes and of the potential substrates.

2. The 'state of the art'

Nowadays different computational approaches are used to predict position of metabolism,[8–19] which can be grouped into QSAR-based, pharmacophore-based, structure-based (docking), reactivity-based, and rule-based methods.

Classical QSAR methods[8] must be applied with particular care, because *they require the same mechanism of action and similar (or homologues) molecules*. This is a strong limitation since the superfamily of P450 cytochrome (CYP) enzymes catalyzes an enormous variety of compounds with a wide variety of oxidative reactions.

Pharmacophore-based substrate methods[9,10] *are strongly training set dependent, and give a static picture of metabolic recognition and reaction* where neither reactivity nor cytochrome active site shape play any role in the overall process. For example, they state that CYP2D6 binds compounds with a basic nitrogen and/or positive charge, and oxidizes atoms at a distance of 5–7 Å from the nitrogen. However, several substrates do exist which have a larger distance between the site of oxidation and the basic nitrogen, e.g. tamoxifen (>10 Å).

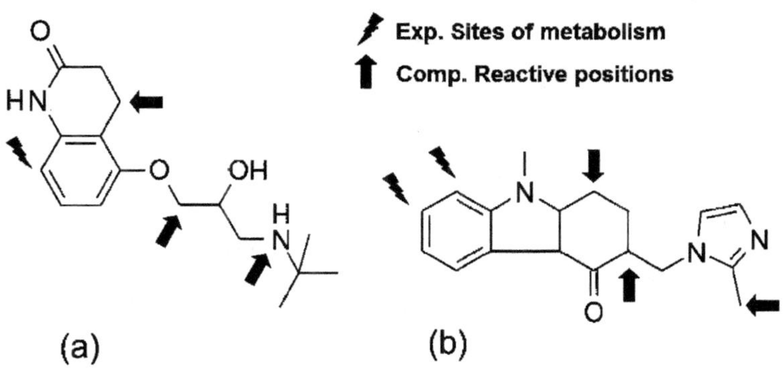

Figure 2. Black arrows indicate the three more reactive positions toward radical abstraction of carteolol (a CYP2D6 substrate on the left) and ondansetron (a CYP3A4 substrate on the right), computed with ab initio method. Grey arrows indicate the experimental sites of metabolism.

Reactivity-based *ab initio* calculations on substrate molecules[11–13] are generally very slow, and *do not take substrate-enzyme recognition and orientation into account*. Figure 2 shows the three more reactive positions in carteolol and ondansetron (CYP2D6 and CYP3A4 substrates, respectively), computed using *ab initio* method.[20] Unfortunately, the only reactive position is at position 8, and the 8-hydroxy-carteolol is the only metabolite formed,[21] while for ondansetron, the only reactive locations are at positions 7 and 8.[22] Chemical reactivity alone, without taking into consideration the orientation of the compound in the reactive-active-site, is far from able to predict the correct site of metabolism.[23]

Rule-based methods[18,19] are based on metabolic transformation rules extracted from the literature and stored in a suitable database, assuming metabolic regularities. These methods *ignore enzymes and 3D structure of compounds*. Rules are assembled with appropriate logic to work in template molecules. As expected much depends on the kind and number of rules, on the training set, on the quality of data from the literature, and on the molecular recognition. Rule-based methods are not human-specific so they produce a relatively high number of possible Phase I reactions. The methods generally over-predict the metabolic transformations giving back hundreds of possible metabolites, and often fail to predict some significant pathways. Moreover, they sometimes fail to identify important minor metabolites.

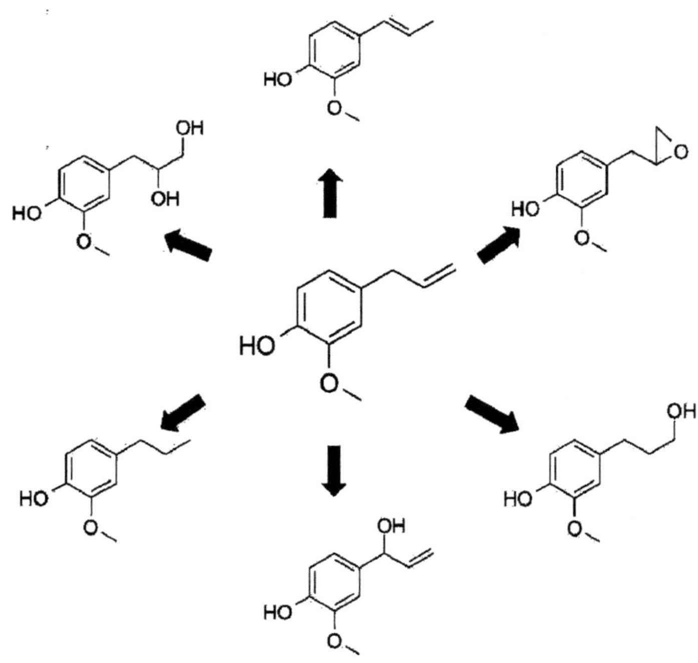

Figure 3. Phase I transformations of eugenol, predicted using a rule-based method. However, experimental findings[24] show that human CYP2D6 mainly catalyzes O-demethylation to produce hydroxychavicol, and this reaction is not included in the transformations above.

Figure 3 shows the possible Phase I transformations of eugenol, predicted using a rule-based method. Each metabolite generated through Phase I reactions may undergo subsequent biotransformations and then Phase II reactions, thus producing hundreds of possible final metabolites. However, experimental findings often show that only one path is populated, or just a few of them. So the question is how are the right ones found?

The knowledge of 3D structural information concerning important human CYPs (such as 2C9, 2D6 and 3A4)[25,26] has revitalized attempts to use docking methods to predict the position of metabolism for drug candidates. However, *these methods are still affected by imprecise scoring functions and by the great flexibility of cytochrome structures*, so they have not as yet improved ability to predict the site of metabolism for xenobiotics. In AstraZeneca Afzelius[27] compared the prediction rate of different methods for a diverse set of compounds, and docking methods were never on the top scoring methods.

3. Results

3.1. THE NEW APPROACH

Metabolism normally only takes place at specific position of a molecular skeleton and, unfortunately, metabolic regularities are exceptions. Researchers have recently focused on developing faster robotic systems and more sensitive analytical metabolite identification tools.[28–32] However, such techniques are particularly resource-demanding tasks, consuming a considerable amount of compound, and cannot be used before synthesis. Moreover, due to the increasing abundance of potential candidates, experimental metabolite identification remains a huge challenge.

We have developed a fast and simple method to answer where is (are) the most likely position(s) of metabolism in a molecular skeleton, and what to do to prevent metabolism. The proposed methodology, called MetaSite (**Site** of **Meta**bolism prediction) involves the calculation of two sets of descriptors, one for the CYP enzyme and one for the potential substrate, respectively representing the chemical fingerprint of the enzyme and the substrate. The set of descriptors used to characterize the CYP enzyme is based on GRID *Flexible* Molecular Interaction Field (GRID-MIFs).[33–35] Flexible Molecular Interaction Fields, reported in Fig. 4, are, in fact, independent of the initial side-chain position of the cytochrome 3D-structure, and better suited to simulate the adaptation of the enzyme to the substrate structure.

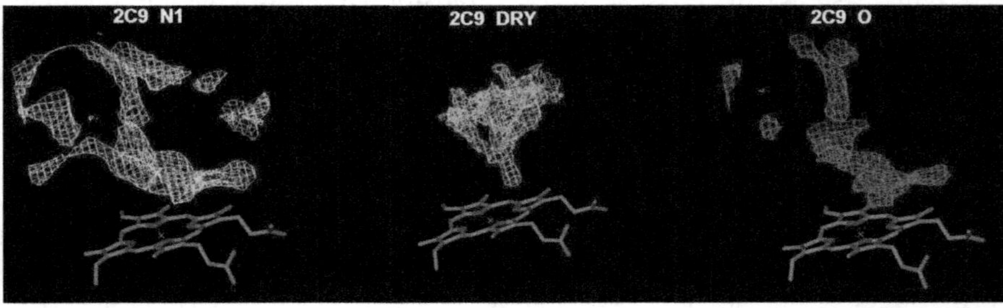

Figure 4. Some of the GRID Molecular Interaction Fields obtained from the same cytochrome are compared.

The descriptors developed to characterize the substrate chemotypes are obtained from GRID probe-pharmacophore recognition. All the substrate atoms are classified into GRID probe categories depending on their hydrophobic, hydrogen-bond donor, acceptor, or charge capabilities. Their distances in the space are then binned and transformed into clustered distances. One set of descriptors is computed for each atom type category: hydrophobic, hydrogen-bond acceptor, hydrogen-bond donor, and charged, which yields a fingerprint for each atom category in the molecule.

The two sets of descriptors are then used to compare the fingerprint of the cytochrome with the fingerprint of the substrate (see Fig. 5).

Three driving forces operate on ligand molecules, substrate of human CYP enzymes. Calculations show that for all the atoms of the test molecules, the probability of being the site of metabolism depends firstly on the enzyme accessibility, called **Ei**, secondly on the chemical reactivity, called **Ri**, and lastly on the reaction mechanism, called **Mi**. Once the three components are calculated, the site of metabolism can be described by a probability function $\mathbf{P}_{SM}$ (Probability for the Site of Metabolism) reported in (Eq. 1), which is correlated to and can be considered to be an approximation of the free energy of the overall process.[36]

$$\mathbf{P}_{SM}\mathrm{i} = \mathbf{E}\mathrm{i} * \mathbf{R}\mathrm{i} * \mathbf{M}\mathrm{i} \qquad (1)$$

where:

- $\mathbf{P}_{SM}$i is the probability of an atom i being the site of metabolism caused by the cyp-heme group
- **Ei** is the accessibility of atom i to the Heme
- **Ri** is the reactivity of atom i in the actual mechanism of reaction
- **Mi** is the relative probability of a reaction mechanism under consideration occurring

Ei is the recognition score between the cyp-protein and the ligand when the ligand is positioned in the cyp-protein and exposes the atom i towards the heme. It depends on the 3D structure, conformation, and chirality of the ligand, and on 3D cyp-protein structure. The **E**i score is proportional to the exposure of atom i to the heme group. Similarly, **R**i is the reactivity of atom i in the appropriate reaction mechanism, and represents the activation energy involved in producing the reactive intermediate. It depends on the

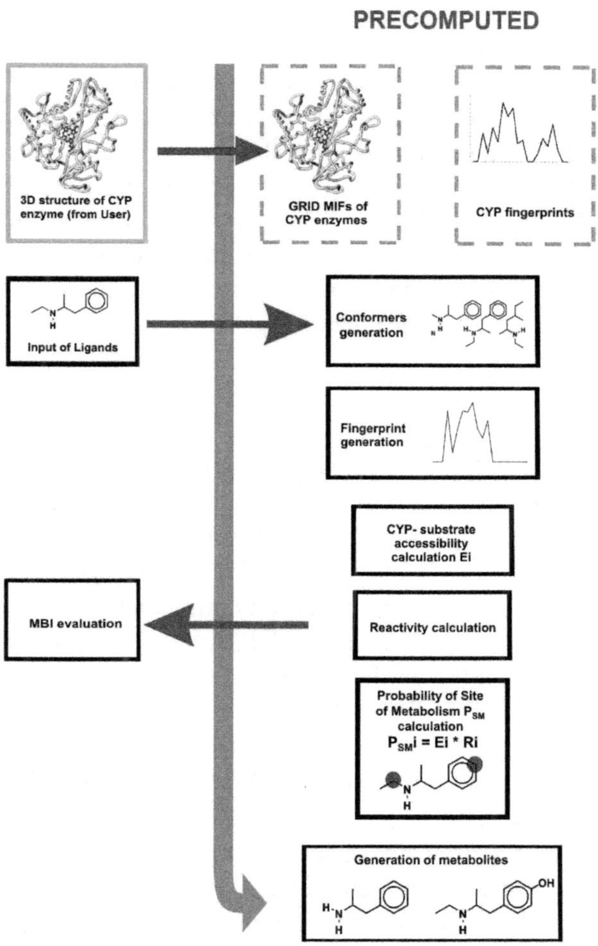

Figure 5. Flow-chart of MetaSite procedure. The GRID-based representations for the main human cytochrome enzymes are pre-computed and stored. However, any cytochrome 3D structure can be imported, with MIF computed on the fly. The ligand pharmacophoric recognition, descriptor handling, and similarity computations are performed automatically once the structure(s) of the compound(s) has been provided. The calculation for chemical reactivity and isoenzyme reactive mechanisms is only performed when exposition has been computed.

3D structure and topology of the ligand. When different reaction mechanisms are possible, **M**i is the relative probability of every mechanism occurring. **M**i can be also considered to be a selectivity factor as it is able to discriminate between reaction mechanisms in different enzymes.

For the same ligand and the same cytochrome, the $\mathbf{P}_{SM}$ function assumes different values for different ligand atoms depending on the **E**i, **R**i and **M**i components. When a ligand atom i is well exposed to the reaction center of the heme (**E**i has a high score), but its reactivity is very low (**R**i reports a very low score), the probability of metabolism in atom i will be very low or zero. Similarly, when a ligand atom i is very reactive in the mechanism considered (**R**i reports a high score), but atom i is not exposed to the reaction center of the heme (**E**i has a very low score), the probability of metabolism in atom i will be close to zero. Therefore, to be the site of metabolism, an atom i should possess significant accessibility and reactivity components related to the heme. However, when two atomic positions i and j show similar values of each single components **E**i or **R**i (or similar value of the product of **E**i * **R**i), then the reaction will favour atom i when mechanism component **M**i is greater for i than for j.

3.2. APPLICATIONS

Figure 6a shows a dianilinophthalimide compound, a potent and selective inhibitor of EGF-receptor kinase.[37] It is well absorbed orally, but it also rapidly metabolizes in man. Drug metabolism studies were carried out to discover the site of metabolism in man. *Para*-hydroxylation on phenylamino moieties was then followed by glucuronylation and excretion. The MetaSite procedure reported above indicates CYP3A4 as the major isoform involved in the oxidation, and the predicted sites of oxidation are reported in the histograms in Fig. 6a. The higher dark bars correspond to the para position in the molecule. In order to prevent drug metabolism, fluorine substituents were placed at the para positions to act as metabolic blockers. The resulting fluoro derivative, also referred to as CGP53353 compound (Fig. 6b), had similar potency but was metabolically stable, in agreement with MetaSite findings.

An inhibitor of intestinal cholesterol absorption (3R)-(3-phenylpropyl)-1,(4S)-bis(4-methoxyphenyl)-2-azetidinone (Fig. 7a) has been demonstrated to lower total plasma cholesterol in man. The potential sites of metabolism in this compound were studied, resulting in a complex metabolite mixture.[38] Further studies confirmed that the mixture was composed of at least four different metabolites, obtained from two different demethylation reactions, plus one benzylic oxidation and one phenyl oxidation (see asterisks in Fig. 7a). Figure 7b reports the probability values for the site of metabolism for all

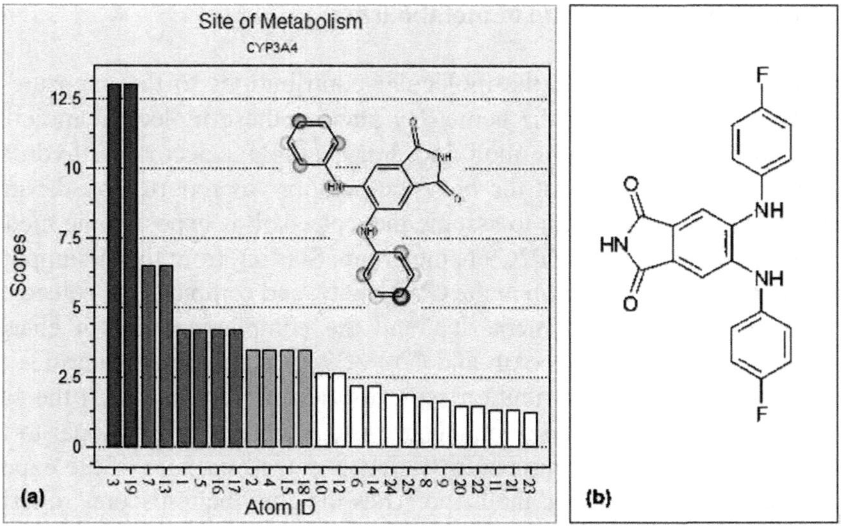

Figure 6. (a) The predicted sites of oxidation are ranked according to probability values, and reported in the histogram. Dark bars highlight the higher probability values that correspond to the para position in the molecule (indicated by dark circles in the 2D structure). (b) The compound CGP53353, a metabolically stable EGF-receptor kinase inhibitor.

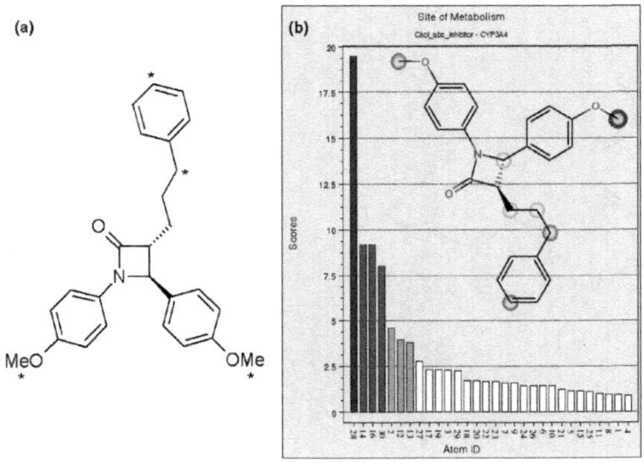

Figure 7. Prediction of the site of metabolism for a cholesterol absorption inhibitor. (a) The asterisks report experimental positions of metabolism. (b) The MetaSite ranking of the probability of metabolism for all the different molecular positions; the first four predicted positions (ranked by probability value) are highlighted.

the atoms of the molecule. The first four ranked positions in Fig. 7b correspond to the circled positions, and exactly match the experimental sites of metabolism reported by asterisks in Fig. 7a. The complex metabolic profile was predicted well, thus showing the great potential impact of this procedure.

4. Contribution to the site of metabolism

MetaSite also can provide the molecular contributions to the exposure of the reactive atom toward the heme. By altering these molecular moieties, the metabolic pattern can be modified. For example, celecoxib is hydroxylated by cytochrome 2C9 at the benzylic position[3] as reported by the gray circle in Fig. 8a. It is logical to assume that celecoxib is exposing the methyl carbon atom toward the CYP2C9 heme group. Starting from this assumption, MetaSite orients the celecoxib in the CYP cavity, and computes the celecoxib-CYP hydrophobic complementarity, and the complementarity of charges and H-bonds between celecoxib and CYP2C9. Such complementarities are then used to assign a contribution score to the different atoms in the substrate. Due to the computation mechanism, the score is proportional to the contribution made by the molecular moieties to the exposure of the experimental reactive atom toward the heme. Thus the contribution scores reported in Fig. 8b highlight the molecular group that influences the hydroxylation reaction most. Chemical modifications of such molecular groups may induce a different site of metabolism. For the purpose under consideration when a derivative of celecoxib substrate is made by replacing the sulfoamine group in celecoxib with a methyl group, the metabolic reaction stability of the new substrate changes from 94% to less than 30%.

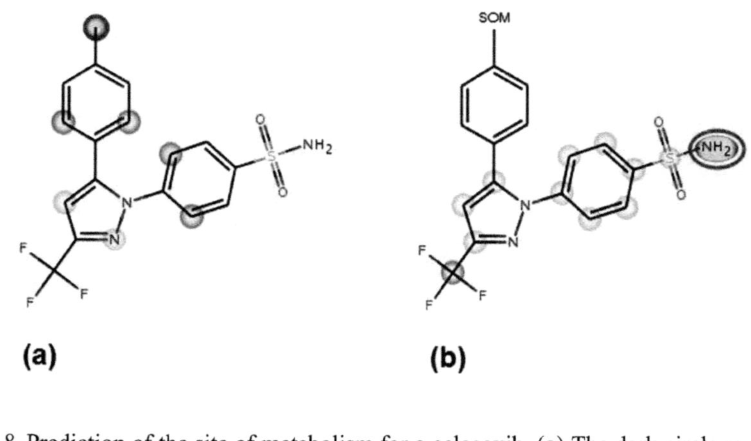

Figure 8. Prediction of the site of metabolism for a celecoxib. (**a**) The dark circle reports the experimental position of metabolism (on benzylic position). (**b**) The MetaSite ranking of the group contribution to metabolism at SOM (Site of Metabolism) position. The SO2NH2 group shows the larger contribution.

5. The influence of the protein structure on the site of metabolism

There are several structures of mammalian (CYP2C5, CYP2B4) and human cytochromes (CYP2C9, CYP2D6 and CYP3A4) that have been crystallized with different ligands. A comparative analysis of these crystal structures can provide an idea about the flexibility of these enzymes and maybe partially explain the broad substrate specificity of these protein structures. Moreover, since a lot of homology models have been created in the past, homology models can also be compared to the crystal structures.

Zhou et al.[39] described for the first time the effect of a crystal structure in the prediction of the site of metabolism when using MetaSite and docking methods. In general docking methods depend strongly on the quality and flexibility of the protein structure and yield different results with respect to a correct prediction of the site of metabolism. The MetaSite procedure is much less depending on the quality and flexibility of the enzyme structure. Interestingly enough, the prediction rate for a docking method with 1w0e crystal structure was worst than predictions based on a homology model developed by De Rienzo et al.[40] The MetaSite methodology, which considers both the protein and the ligand as flexible entity, is less dependent on the protein structure.

Afzelius et al.[27] described the prediction rate for the site of metabolism comparing different computational methods for a diverse set of compounds based on reactivity (bond order computations biased for surface accessible area), knowledge-based approach like SPORCalc, the MetaSite method, a docking method based on the Glide software, and finally, a prediction made by scientists with more than 20 years of experience on metabolite identification. MetaSite was always the best method overall. Moreover, the authors described the effect of the protein structure on the prediction rate depending on several crystal structures for CYP3A4 (2j0c and 2j0d).

Another study has been performed[41] to compare the MetaSite prediction rates based on different CYP crystal structures and the homology models for CYP2C9 and other CYP Xray structures (Fig. 9). The protein structures were imported into MetaSite and the three top ranked solutions were collected for each case using a diverse set of substrates. The combination columns were obtained as a consensus prediction. The 2C9 homology model and the 1r90 Xray structure seems the best structure for site of metabolism prediction.

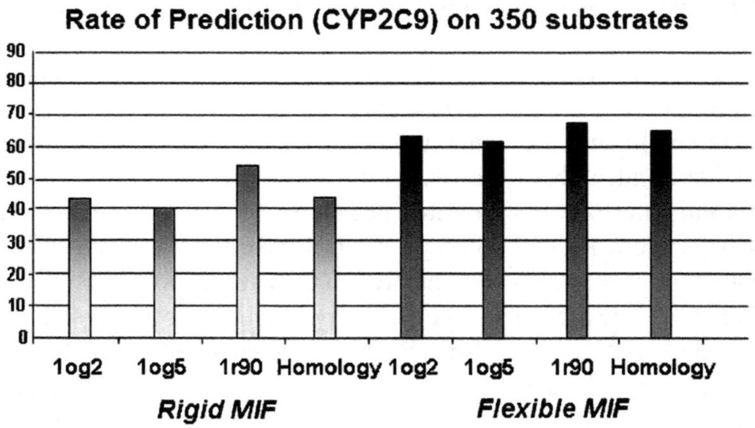

Figure 9. Influence of the protein 3D structure on the rate of prediction for CYP2C9 with 350 substrates. It is important to point out that the correct predictions for the different protein structures are not the same.

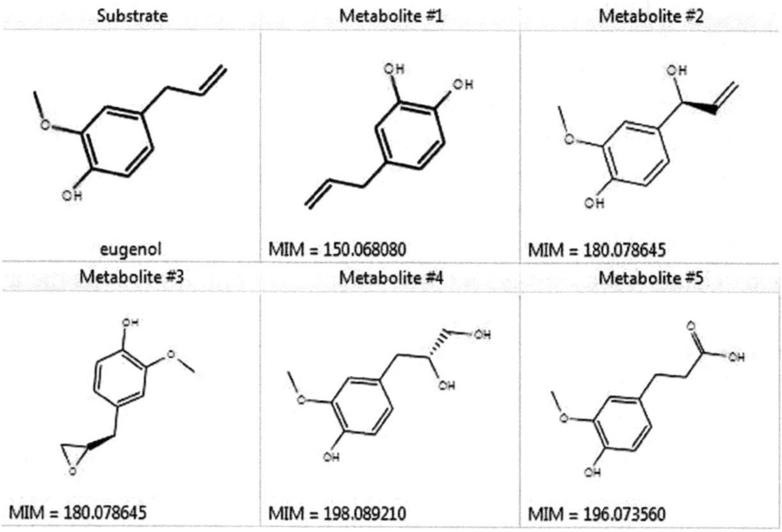

Figure 10. Phase I transformations of eugenol, predicted using MetaSite. The first metabolite correspond to hydroxychavicol, the main metabolite structure. Compare this transformations with those of Fig. 3. Please note that the results are obtained with NO TRAINIG data set, since MetaSite is NOT a training set dependent method.

6. The computation of metabolic pathway

Once the site of metabolism probability is computed, the generation of the main metabolite structures is not difficult. The metabolite are sorted according to the computed score obtained in the hot spot prediction phase. Therefore, the metabolite ranking reflects the probability of site of metabolism. This is a great advantage respect to other procedures, which produce metabolites based on other criteria, but without associating to them an absolute ranking scheme.

7. Conclusions

A methodology has been developed to predict the site of metabolism, the contribution to the site of metabolism, and the ligand-cytochrome complementarity for substrates of the most important human cytochromes. On average, in about 85% of cases the method predicted the correct site of metabolism within the first two choices in the ranking list.

The methodology works for the most important human cytochromes, but can be automatically applied to all the cytochromes whose the 3D structure is known.

It is important to stress that the method highlighted here requires neither training nor docking procedures and associated scoring functions, nor 2D or 3D QSAR models. The methodology does not use any training set, or supervised or unsupervised technique. In contrast, the method relies on flexible molecular interaction fields generated by the GRID force field on the CYP homology modeling structures that were treated and filtered in order to extract the most relevant information. The fully automated computational procedure is a valuable new tool in early ADME-Tox, where drug-safety and metabolic profile patterns must be evaluated in order to enhance and streamline the process of developing new drug candidates.

References

1. R. Iyer and D. Zhang, in *Drug metabolism in drug design and development,* D. Zhang, M. Zhu, W.G. Humphreys Eds, Wiley 2008, p. 267.
2. M. Zhu, W. Zhao, W. G. Humphreys, in *Drug metabolism in drug design and development*, D. Zhang, M. Zhu, W.G. Humphreys Eds, Wiley 2008, pp. 287–313.
3. M. Ablström, M. Ridderström, K. Luthman, I. Zamora, *J. Med Chem.* **50**(18), 4444–4452 (2007).

4. M. Rowley, D.J. Hallett, S. Goodacre, C. Moyes, J. Crawforth, T.J. Sparey, Sm. Patel, R. Marwood, Sh. Patel, S. Thomas, L. Hitzel, D. O'Connor, N. Szeto, J.L. Castro, P.H. Hutson, A.M. MacLeod, *J. Med. Chem.* **44**(10), 1603–1614 (2001).
5. I. Zamora, L. Afzelius, G. Cruciani, *J. Med. Chem.* **46**, 2313–2324 (2003).
6. G. Berellini, G. Cruciani, R. Mannhold, *J. Med. Chem.* **48**(13), 4389–4399 (2005).
7. G. Cruciani, E. Carosati, B. De Boeck, K. Ethirajulu, C. Makie, T. Howe, R. Vianello, *J. Med. Chem.* **48**(7), 2445–2456 (2005).
8. Hasselgren-Arnby C, Smith J, Glen RC, and Boyer S (2005) SPORCalc-fingerprint based probabilistic scoring of metabolic sites, in *The 7th International Conference on Chemical Structures*. 5–9 Jun, 2005; Abstract C-2; Noordwijkerhout, The Netherlands.
9. M.J. De Groot, M. Ackland, V. Horne, A. Alexander, J. Barry, *J. Med. Chem.* **42**, 4062–4070 (1999).
10. B.C. Jones, G. Hawksworth, V.A. Horne, A. Newlands, J. Morsman, M.S. Tute, D.A. Smith, *Drug Metab. Disp.* **24**, 260–266 (1996).
11. S.B. Singh, L.Q. Shen, M.J. Walker, R. Sheridan, *J. Med. Chem.* **46**, 1330–1336 (2003).
12. H. Chen, M. de Groot, N. Vermulen, R.P. Hanzlik, *J. Org. Chem.* **62**, 8227–8230 (1997).
13. S.P. Visser, F. Ogliaro, P.K. Sharma, S. Shaik, *J. Am. Chem. Soc.* **124**, 11809–11826 (2002).
14. K.R. Korzekwa, J. Grogan, S. DeVito, J.P. Jones, *Adv. Expl. Med. Biol.* **38**, 361–369 (1996).
15. D.F. Lewis, M. Dickins, P.J. Eddershaw, M.H. Tarbit, P.S. Goldfarb, *Drug Metab. Drug Interac.* **15**, 1–49 (1999).
16. A. Mancy, P. Broto, S. Dijols, P.M. Dansette, D. Mansuy, *Biochemistry* **34**, 10365–10375 (1995).
17. M. Riddestrom, I. Zamora, O. Fjäström, T.B. Andersson, *J. Med. Chem.* **44**, 4072–4081 (2001).
18. F. Darvas, S. Marokhazi, P. Kormos, G. Kulkarmi, H. Kalasz, A. Papp, in: *Drug Metabolism*; Erhardt, P.W. Ed., Blackwell Science, 1999, pp 237–270.
19. B. Testa, A.L. Balmat, A. Long, P. Judson, *Chem Biodivers.* **2**, 872–885 (2005).
20. Open-shell radicals were optimized at AM1 semi-empirical level. Single point energy evaluations were performed by DFT at the B3LYP/6-311G** level of theory since correlation between experimental and calculated radical stabilities resulted in reasonable agreement for this level of theory.
21. S. Kudo, M. Uchida, M. Odomi, *Eur. J. Clin. Pharm.* **52**, 479–485 (1997).
22. J.F. Pritchard, *Semin. Oncol.* **19**, 9–15 (1992).
23. C. de Graaf, N.P.E. Vermeulen, K.A. Feenstra, *J. Med. Chem.* **48**, 2725–2755 (2005).
24. S. Katsuhisa, I. Yuji, O. Shinji, H. Yusuke, K. Shosuke, *Mutat. Res.* **565**, 35–44 (2004).
25. M.R. Wester, J.K. Yano, G.A. Schoch, K.J. Griffin, C.D. Stout, E.F. Johnson, http://www.pdb.org, 2004, 1R9O entry.
26. J.K. Yano, M.R. Wester, G.A. Schoch, K.J. Griffin, C.D. Stout, E.F. Johnson, http://www.pdb.org, 2004, 1TQN entry
27. L. Afzelius, C.H. Arnby, A. Broo, L. Carlsson, C. Isaksson, U. Jurva, B. Kjellander, K. Kolmodin, K. Nilsson, F. Raubacher, L. Weildolf, Drug Metab. Rev. **39**(1), 61–86 (2007).
28. S.R. Thomas, U. Gerhard, *J. Mass Spectrom.* **39**, 942–948 (2004).
29. E. Kantharaj, A. Tuytelaars, P. Proost, Z. Ongel, H.P. Assouw, R.A. Gilissen, *Rapid Commun. Mass Sp.* **17**, 2661–2668 (2003).
30. R. Kostiainen, T. Kotiano, T. Kuurama, S.J. Auriola, *Mass. Spectrom.* **38**, 357–372 (2003).

31. O. Corcoran, M. Spraul, *Drug Discov. Today* **8**, 624–631 (2003).
32. A.E.F. Nassar, R. E. Talaat, *Drug Discov. Today* **9**, 317–327 (2004).
33. P.J. Goodford, *J. Med. Chem.* **28**, 849–857 (1985).
34. E. Carosati, S. Sciabola, G. Cruciani, *J. Med. Chem.* **47**, 5114–5125 (2004).
35. P.J. Goodford, in *Rational Molecular Design in Drug Research*, Alfred Benzon Symposium 42, Liljefors, T., Jorgensen, F.S., Krogsgaard-Larsen, P. Eds.; Munkgaard, Copenhagen 1998, pp. 215–230.
36. MetaSite ver. 3.0, Molecular Discovery Ltd, 2008 (http://www.moldiscovery.com)
37. U. Trinks, E. Buchdunger, P. Furet, W. Kump, H. Mett, T. Meyer, M. Muller, U. Regenass, G. Rihs, N. Lydon, *J. Med. Chem.* **37**, 1015–1027 (1994).
38. S.B. Rosemblum, T. Huynh, A. Afonso, H.R. Davis, N. Yumibe, J.W. Clader, D. Burnett, *J. Med. Chem.* **41**, 973–980 (1998).
39. D. Zhou, A. Afzelius, S.W. Grimm, T.B. Andersson, R.J. Zauhar, I. Zamora, *Drug Metab. Dispos.* **34**, 976–983 (2006).
40. F. De Rienzo, F. Fanelli, M.C. Menziani, P.G. De Benedetti. *J. Comp.-Aided Mol. Des.* **14**, 93–116 (2000).
41. I. Zamora, in *Antitargets*, R.J. Vaz and T. Klabunde Eds, Wiley-VCH 2008, pp. 247–255.

THE FOUNDATIONS OF PROTEIN–LIGAND INTERACTION

GERHARD KLEBE*
Institute of Pharmaceutical Chemistry, University of Marburg, Marbacher Weg 6, D35032 Marburg, Germany

Abstract. For the specific design of a drug we must first answer the question: How does a drug achieve its activity? An active ingredient must, in order to develop its action, bind to a particular target molecule in the body. Usually this is a protein, but also nucleic acids in the form of RNA and DNA can be target structures for active agents. The most important condition for binding is at first that the active agent exhibits the correct size and shape in order to optimally fit into a cavity exposed to the surface of the protein, the "**binding pocket**". It is further necessary for the surface properties of the ligand and protein to be mutually compatible to form specific interactions. In 1894 Emil Fischer compared the exact fit of a substrate for the catalytic centre of an enzyme with the picture of a "lock-and-key". Paul Ehrlich coined in 1913 "*Corpora non agunt nisi fixata*", literally "bodies do not work when they are not bound". He wanted to imply that active agents that are meant to kill bacteria or parasites must be "fixed" by them, i.e. linked to their structures. Both concepts form the starting point for any rational concept in the development of active pharmaceutical ingredients. In many respects they still apply today. A drug must, after being administered, reach its target and interact with a **biological macromolecule**. Specific agents have a large affinity and sufficient selectivity to bind to the macromolecule's active site. This is the only way they can develop the desired biological activity without side-effects.

Keywords: Drug design, binding pocket, protein-ligand interaction, gibbs free energy, hydrogen-bonding networks, ionic interactions, charge-assisted hydrogen bond, hydrophobic interactions

---

* To whom correspondence should be addressed. Gerhard Klebe, Institute of Pharmaceutical Chemistry, University of Marburg, Marbacher Weg 6, D35032 Marburg, Germany, Fax +49 6421 282 8994; e-mail: klebe@mailer.uni-marburg.de

J.L. Sussman and P. Spadon (eds.), *From Molecules to Medicines*,

1. The binding constant Ki describes the strength of the protein–ligand interaction

The binding of a ligand to its target protein can be measured. The **Binding Constant K_i** Eq. (1) may be regarded as a characteristic binding quantity. To be precise, it is a dissociation constant K_D, and its reciprocal is the association constant K_A. The inhibition constant K_i of enzymes is determined in an assay. Although they do not describe exactly the same, these values are generally used equivalently. In the following only the abbreviation K_i will be used. The binding constant describes the strength of the interaction between protein and ligand. It is a thermodynamic state function and reflects how much of the ligand is, on the average, bound to the protein. The following law of mass action may be derived:

$$K_i = \frac{[\text{Ligand}]*[\text{Protein}]}{[\text{Ligand}*\text{Protein}]} \tag{1}$$

K_i has the dimension of a concentration with the unit of mol/l (M). The smaller the value of K_i, the stronger does the ligand bind to the protein. If the concentration of the ligand is significantly less than K_i, then only a small percentage of ligand molecules will be bound to protein molecules. A biological effect such as the inhibition of an enzyme cannot be observed. If the ligand concentration corresponds to K_i, then half of all the protein molecules present are bound by ligand molecules. The binding constant may be transformed into the Gibbs free energy of binding ΔG by using the following relationship from thermodynamics Eq. (2)

$$\Delta G = RT \ln K_i \tag{2}$$

In Eq. (2), R is the gas constant and T the absolute temperature in Kelvin. A binding constant of $K_i = 10^{-9}$ M = 1 nM, a respectable value for an active agent, corresponds at body temperature to a Gibbs free binding energy of -53.4 kJ/mol. A change of one order of magnitude for K_i means a change in the Gibbs free binding energy of 5.9 kJ/mol or 1.4 kcal/mol, respectively.

Often, instead of the K_i value, a so-called IC_{50} value is determined. This quantity indicates at which ligand concentration the activity of the protein (usually enzyme) has been decreased by 50%. In contrast to the K_i value, the IC_{50} value depends on the concentration of the enzyme. Experience has shown that both values run approximately parallel so that IC_{50} values, which are easier to determine, are very well suited for the characterisation of a ligand in comparison to other structures.

Why is the **Gibbs free energy** (or **free enthalpy)** used here to describe the energy relationships during complex formation? In chemistry and biology, processes take place in open systems under atmospheric pressure. Because the volume of the environment is so enormously large, it can be assumed that the external pressure remains unchanged, even for processes that involve the development of gases. These are therefore regarded as processes under constant pressure conditions. Nevertheless, a gas that is formed during a reaction must first create its own volume in competition with the surrounding air particles. It must perform work. This so-called **volume work** will diminish the maximally possible work to be performed by a system ("**inner energy**" ΔU). The energy diminished by the volume work is called the **enthalpy** (ΔH). This is therefore the energy converted during a process corrected by the amount of volume work. The release of enthalpy is not yet the complete answer as to why a certain process such as the formation of a protein–ligand complex **spontaneously** takes place. Let us take a hot and a cold block of metal and let them contact each other. Everyone knows that heat will flow spontaneously from the warm to the cold block. The reverse is not observed although the energy content of the system investigated during this process would remain unchanged. Why then does energy flow spontaneously from the hot to the cold object and not the other way round? This has something to do with the tendency of all processes in nature to distribute energy evenly. In the hot metal block, the metal atoms vibrate strongly around their resting positions. This is why the block is hot. Several vibrational degrees of freedom are strongly activated. If the hot and cold blocks are now brought into contact, the vibrations are transmitted. Finally, the metal atoms in both blocks vibrate about their resting positions, but on average not so violently as the atoms previously in the hot block. The energy content has indeed remained constant, but is now distributed over many more degrees of freedom. We can say that the system has proceeded into a more **disordered state** (many atoms now vibrate on average more strongly). This happens in all processes that take place spontaneously. The value used to quantify this uniform or randomly disordered distribution is the **entropy** S. In order to correctly describe the processes involved in the formation of a protein-ligand complex Eq. (3), we need more than just the enthalpy (ΔH) exchanged between the binding partners during the process. It must be regarded by how much the energy distribution among the degrees of freedom changes and whether the system migrates during this process to a state of enhanced disorder. We therefore use here the Gibbs free energy (ΔG), because this quantity takes not only the energy balance of the process into consideration. It also considers changes in entropy ($-T\Delta S$) due to the spontaneous distribution of energy among the system's degrees of freedom.

$$\Delta G = \Delta H - T\Delta S \tag{3}$$

ΔG is comprised of an enthalpic ΔH and an entropic $-T\Delta S$ component Eq. (3). The entropic component is weighted with the temperature. It is of great importance whether the entropy in a system changes at low temperatures where all particles are already predominantly ordered, or at high temperatures where the disorder is already very high. Because of the negative sign, an increase in entropy means a decrease in ΔG and consequently an increase in the binding affinity.

2. Important types of protein-ligand interactions

Organic molecules can bind to proteins not only by forming a chemical bond between ligand and protein but also via non-covalent interactions. Omeprazole, for example, reacts chemically with the protein and forms a covalent bond. In the following we want to restrict ourselves to just those ligands that bind to proteins via non-covalent interactions. It will be convenient for the following discussions to classify protein-ligand interactions into various categories. The different types of interactions are summarised in Fig. 1.

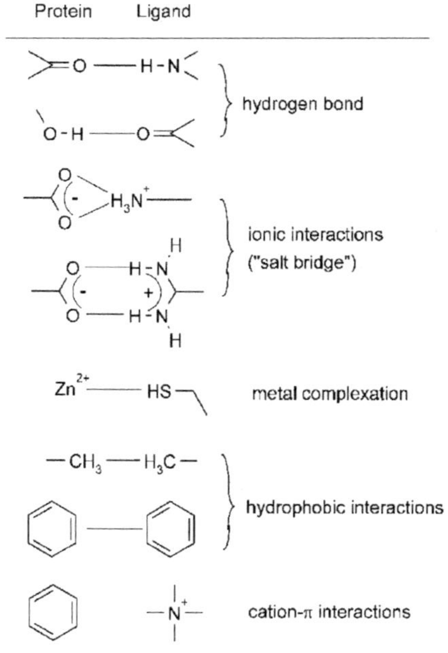

Figure 1. Frequently occurring protein–ligand interactions. Important polar interactions are hydrogen bonds and ionic interactions. Metalloproteases contain as cofactor zinc ions. Their interactions with ligands often contribute significantly to binding affinity. Apolar parts of the protein and the ligand contribute to the binding by hydrophobic interactions. Because of the exceptional electron distribution in aromatic ring systems the interaction between these unsaturated systems is especially large.

Very often, **hydrogen bonds** (in short H-bonds) between protein and ligand can be observed. The partner carrying the proton, in biological systems usually an >NH or –OH group, is called a **hydrogen-bond donor**. The counter group is an electronegative atom with a partial negative charge and is called a **hydrogen-bond acceptor**. Hydrogen bond acceptors are, for example, oxygen and nitrogen atoms. Hydrogen bonds are predominantly of electrostatic nature. They achieve their extraordinary strength because the proton from the donor group is bonded to a strongly electronegative atom and thereby the electron density is shifted from the proton to the neighbouring atom. The sphere of influence of the hydrogen atom becomes virtually smaller. This in turn allows the acceptor to approach closer to the proton in the H-bond than the sum of the van der Waals radii would apparently permit. The electrostatic attraction between the partners therefore becomes larger. The geometry of an H-bond is shown in Fig. 2. A hydrogen bond is characterised by a pronounced distance and angular dependency. This dependency is directional and its geometry is defined within narrow limits.

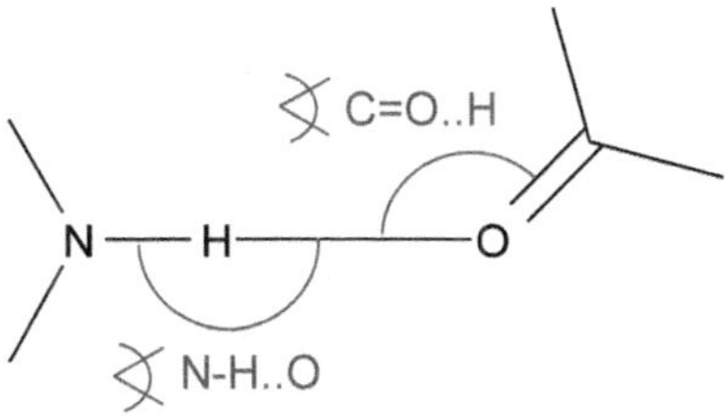

Figure 2. The geometry of a hydrogen bond. The atoms N, H and O assume a nearly linear arrangement to each other. The distance N...O falls between 2.8 and 3.2 Å. The angle N-H...O is in most cases larger than 150°. A larger range is observed for the angle C=O...H. This typically lies between 100° and 180°.

It is often observed that charged groups of the ligand bind to oppositely charged groups of the protein. Such **ionic interactions** (also known as salt bridges) are especially strong due to the electrostatic attraction of the charges at a distance of 2.7–3.0 Å to each other. Often a hydrogen bond is superimposed onto an ionic interaction. This is called a **charge-assisted hydrogen bond**. We will see that in many protein-ligand complexes the association is determined to a significant degree by such ionic interactions. A few proteins contain metal ions as cofactors, for example Zn^{2+} in metalloprotease. In these structures, an attractive interaction between the metal ion and the opposite charge in the ligand is often a decisive contribution to affinity. Further, there are a few groups that are particularly suitable for complexing a transition metal. Examples are thiols RSH, hydroxamates RCONHOH, acid groups and many nitrogen containing heterocycles.

Hydrophobic interactions arise from the close contact of apolar amino acid side chains of the protein and lipophilic groups of the ligand. Lipophilic groups are not only aliphatic or aromatic carbon residues but also halogen substituents (e.g. chlorine) and many heterocycles such as, for example, thiophene and furan (Fig. 3). All areas of a protein and ligand that cannot themselves form H-bonds or any other polar interactions are considered as lipophilic areas of the surface. Hydrophobic interactions are, in comparison to hydrogen bonds, not directional. It is unimportant in which relative orientation the lipophilic groups orient towards each other. An exception are interactions between aromatic portions for which there do exist a preferred relative orientation.

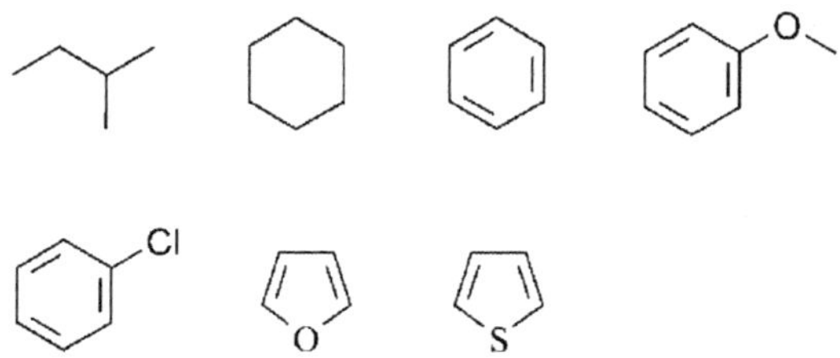

Figure 3. Typical lipophilic groups in ligands are aliphatic and aromatic hydrocarbons, halogen substituents or apolar heterocycles such as furan and thiophene.

It has been shown that hydrophobic interactions often provide for ligands with large lipophilic groups a very important contribution to the binding affinity. The influence of directly attracting forces between the lipophilic groups is, however, faint. Instead, the hydrophobic interaction is caused mainly by the displacement or, to put it more precisely, the **discharge of water molecules** from the lipophilic environment of the binding pocket. Furthermore, the ligand with its lipophilic substituents leaves the aqueous solution around the protein. The "solvent cave" in the water which hosted the ligand, collapses. This step is also associated with changes in the Gibbs free energy. The role of the water molecules will be discussed in Section 4.

One further important interaction must be mentioned here. Quaternary amines evidently bind readily in binding pockets that are formed by aromatic side chains of the protein. This contact is largely based upon the polarisation interaction between the positive charge and the electron system of the aromatic moieties.

3. The strength of protein-ligand interactions

In order to assess the strength of protein-ligand interactions, it is reasonable to first consider the non-covalent interactions between isolated small molecules. Information on this is available both from quantum-mechanical calculations and from spectroscopic studies. Molecular pairs in the gas phase may be investigated by these means. The observed association energies of the molecules provide an impression of the strength of the direct interactions. The influence of effects that originate from the liberation of solvent waters (**desolvation**) are of course missing in these experiments. Some data are presented in Table 1.

The results show that electrostatic interactions are the dominating energetic factor. The interaction between an anion and a cation in vacuum amounts to more than 400 kJ/mol. This corresponds to the strength of a covalent bond! In comparison to the typical protein-ligand interaction, as given in Sections 6 and 7, this is an enormous amount. In the gas phase, the binding energy of an ion pair is therefore considerably larger than the typical strength of protein-ligand interactions in water. Two water molecules bind to each other with 22 kJ/mol. This interaction is also predominantly electrostatic in nature, whereby the quite large dipole moment is responsible for the strong binding. Interactions between small, apolar molecules are considerably weaker. Two methane molecules interact with approx. 2 kJ/mol. This is less than 10% of the H_2O...H_2O interaction. Correspondingly, methane boils at 90 K, whereas water is liquid at room temperature. The direct interaction between polar groups is therefore orders of magnitude larger than that between apolar groups.

TABLE 1. Experimentally determined or quantum-chemically calculated association energies in the gas phase.

Dimer	Binding energy (kJ/mol)
CH_4 … CH_4	–2.0
CH_6 … CH_6	–10
H_2O … H_2O	–22
NH_3 … NH_3	–18
Na^+ … H_2O	–90
NH_4^+ … $COOH^-$ Na^+ … Cl^-	<–400

4. Blame it all to water!

The data presented in the previous paragraph could give the impression that protein-ligand interactions are determined mainly by H-bonds and ionic interactions. It is therefore even more surprising to realise that the acetate ion CH_3COO^- does not form dimers with the guanidinium ion $H_2NC(=NH_2^+)NH_2^-$. Likewise amides practically never associate in water although the hydrogen bond between two amide groups in protein structures often occurs. How can this be? The answer is: water is to blame for everything!

All biochemical reactions take place in water and this is the only reason that they occur. The binding of a ligand to a protein takes place in **aqueous solution**. The "empty" binding pocket of the protein is first filled with water molecules. Some of the water molecules form hydrogen bonds to the protein and adopt an energetically favourable interaction pattern. Other water molecules make contact with the lipophilic areas of the protein's surface and cannot form any perfect **hydrogen-bonding networks**. The ligand is also solvated. When it diffuses into the binding pocket, it displaces the water molecules found there. In addition, it has to strip off its solvation shell. At the same time, the "solvent cave" in which the ligand resided in the water collapses. Consequently, not only direct interactions between protein and ligand are formed, but also numerous H-bonds to water molecules are ruptured.

We want to look more precisely at the formation of a hydrogen bond and lipophilic contacts between protein and ligand. Both processes are shown in Fig. 4. How is an H-bond between protein and ligand formed? Let us assume that the polar groups of both partners are solvated.

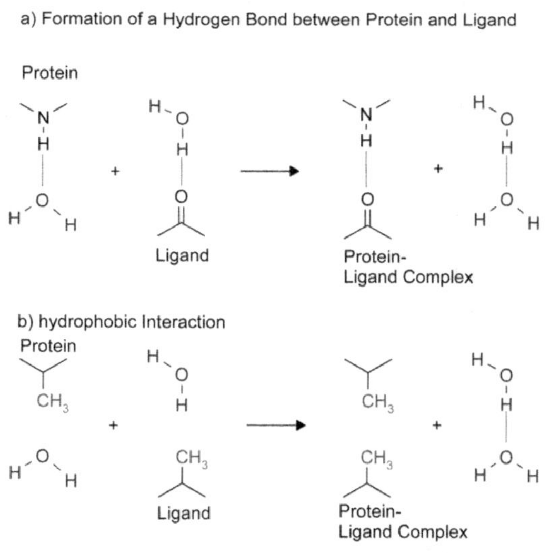

Figure 4. Influence of water molecules on the strength of protein-ligand interactions. (**a**) During the formation of an H-bond between the protein and ligand, the water molecules must be displaced. They form themselves H-bonds to the protein and ligand, respectively. The balance of H-bonds, i.e. the number of H-bonds before and after the forming a hydrophobic contact, remains unchanged. (**b**) During the formation of a hydrophobic contact, water molecules are liberated from an environment that is unfavourable for them to the bulk water. The number of H-bonds increases.

Then, for the formation of an H-bond between the two, at least two water molecules must be displaced. These may in turn form H-bonds with other water molecules. Consequently, exactly the same number of H-bonds is broken as is newly created. The total number of H-bonds therefore remains constant! The gain in free binding enthalpy is determined by the relative strength of the various H-bonds as well as by the entropic contributions that relate to the changes in the degree of order of the system (Section 5). It is difficult to quantify the net contribution to the Gibbs free energy. If the ligand succeeds in forming more hydrogen bonds to the protein than it could achieve in its solvent shell, this will result in a very strong binding. This is particularly the case if, in the binding pockets of the protein, the groups forming the polar H-bonds are orientated such that the water molecules alone cannot fully satisfy all these interactions. In contrast, the ligand may accomplish this through an optimal arrangement of its donor and acceptor groups.

Furthermore the formation of a hydrophobic contact leads to the liberation of water molecules from the binding pocket. They now can form H-bonds amongst themselves (Fig. 4). Because H-bonds neither to the ligand nor to the protein were previously possible, the total number of H-bonds now increases. The enhanced mobility of the water molecules released to the bulk increases the disorder and therefore boosts the entropy, which is thermodynamically favourable for the Gibbs free energy of binding ΔG. Recent results have shown that binding pockets must not always be solvated by a closely packed cluster of water molecules. In particular, narrow hydrophobic pockets are not perfectly solvated. This has consequences for the energy balance during binding because the displacement of the water molecules is decisive for the hydrophobic interactions.

5. Entropic contributions to protein–ligand interactions

When considering the strength of protein-ligand interactions, not only the energetic contributions but also the entropic components must be taken into account. **Entropy** S is, as already described, a measure of the order (or disorder) of a system. It allows to estimate over how many degrees of freedom a certain energy content is distributed over the system. One degree of freedom can mean, for example, a certain vibration of the system or a rotation of two single groups around each other. A highly ordered system, which contains only a few degrees of freedom, has a low entropy, increasing the disorder increases the entropy and consequently lowers the Gibbs free energy G.

At room or body temperatures, protein and ligand can move in all spatial directions. In addition, the water environment is, of course, also moveable, water molecules diffuse back and forth. Some of them are locally arrested over long periods of time, namely when they are bound by several H-bonds to the protein. Such water molecules may be identified by X-ray crystallography of the protein. A local fixation of the molecule is, however, entropically unfavourable. Other water molecules are free to move and therefore cannot be detected in an X-ray structure determination. Such water molecules exist in an entropically favourable state, because their $-T\Delta S$ contribution is more favorable than for a spatially fixed water molecule.

The hydrophobic protein-ligand interaction is, as we have already seen, of largely entropic nature. Individual water molecules are displaced out of the binding pocket and released into the surrounding bulk water. Accordingly, the entropic contribution to the protein-ligand interaction is based not on direct interactions, but rather on the change of the number of degrees of freedom in the system protein – ligand – water through the association of the ligand with the protein. The more water molecules are released from the hydrophobic environment, the larger is the contribution to the binding affinity. The number of water molecules released is, to a first approximation, directly proportional to the size of the hydrophobic surface that is no longer available to water after binding of the ligand to the protein, i.e. it can be understood as "buried". This surface contribution is therefore often used as a benchmark value for estimating the entropic portion to binding.

In addition to the release of water molecules, there are other entropic contributions to the binding energy. The association of the ligand with the protein leads to a loss of translational and rotational degrees of freedom and consequently to a loss of entropy. Prior to complex formation protein and ligand move freely and independently of each other. They have each three degrees of translational and rotational freedom. After binding, the protein and the ligand diffuse and rotate as a complex, which means that three translational and rotational degrees of freedom are lost. A freely mobile, flexible ligand can furthermore assume various conformations which is therefore entropically favourable. The ligand that is bound to the protein is restricted in its conformational degrees of freedom to one or just a few conformations that fit into the binding pocket of the protein. The ligand finds itself in an entropically unfavourable state. Various enthalpic and entropic contributions to binding are summarised in Fig. 5.

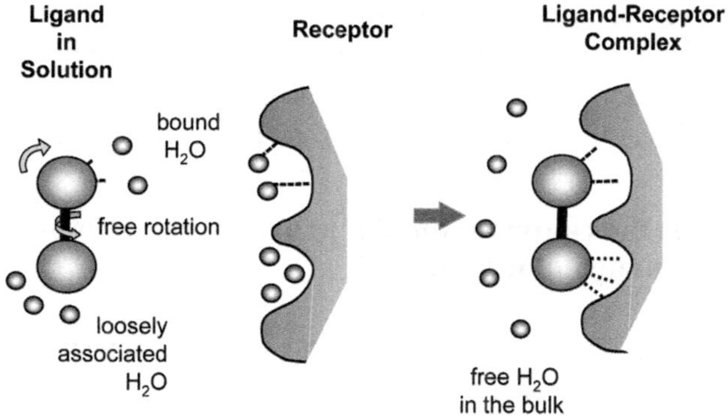

Figure 5. Illustration of the thermodynamic contributions to the Gibbs free binding energy ΔG. Before binding, the ligand moves freely. It possesses a certain translational and rotational entropy. Further, the ligand is mostly flexible and adopts various conformations. Protein and ligand are solvated, whereby H-bonds to water molecules are formed. A few water molecules find themselves in loose contact to the protein or the ligand without forming H-bonds. During binding, degrees of translational and rotational freedom are lost. This results in a lowering of the binding entropy. Furthermore, the protein and ligand must release part of their solvation shell, a likewise unfavourable process for binding. Ligand binding leads to the formation of direct interactions to the protein and releases water molecules from the binding site to the bulk. These are both contributions that most likely favour binding. H-bonds are represented as dashed lines, hydrophobic contacts as dotted lines.

One would first assume that the entropic contribution $-T\Delta S$ to ΔG is positive and the enthalpic contribution ΔH negative. In fact such an **enthalpy driven binding** is frequently observed. However, there are also many cases known, mainly for large lipophilic ligands, in which the binding is entropy driven. This means that the binding of the ligand is enthalpically unfavourable, this effect is, however, overcompensated for by a marked increase of entropy, in consequence ΔG becomes in total negative. The gain in entropy arises, as already mentioned, firstly by the liberation of water molecules. This is, however, not the only contribution to entropy that changes upon ligand binding. The protein also changes. For example, many side chains in proteins are distributed over several conformational states. When accommodating a ligand, this distribution may change. Depending on the total balance, the entropy can increase or decrease. The same applies to the rotation of side chains, mainly methyl groups. Changes in their rotational behaviour influence likewise the total entropy of the ligand binding process. The total picture can become so entangled that during binding, some areas of the protein migrate to more ordered, others to less ordered states. The entropic contributions therefore partially compensate each other. It has often been assumed that the entropic changes upon binding remain constant within a series of very similar ligands. In this case it would not be necessary to be concerned with such contributions while comparing relative differences of these ligands amongst themselves. Unfortunately, this simple picture has proved to be a fallacy. A case in point will be presented in Section 8.

6. How large is the contribution of a hydrogen bond to the strength of the protein–ligand interaction?

Discussing protein–ligand interactions, the inherent question arises as to how large the contribution of a particular hydrogen bond to the binding affinity actually is. This question may be answered experimentally when two protein-ligand complexes are compared to each other that only differ from one another by one hydrogen bond. Such a comparison is possible, for example, by using protein mutants in which one single amino acid that forms an H-bond to the ligand is replaced by another residue not capable to form this interaction. An elegant experiment was performed by Alan Fersht for the protein tyrosyl-tRNA synthetase in complex with the substrate tyrosyl-adenylate (Fig. 6). Numerous H-bonds are formed between the protein and substrate, for example between the phenolic OH groups of the tyrosine 34 and the substrate. The mutant Tyr 34 → Phe in which the tyrosine is replaced by apolar phenylalanine was prepared and the binding of the substrate to the protein mutant tested. Due to this mutation, the binding was weakened by 2 kJ/mol. Similarly, other mutants have been investigated. The loss of an uncharged H-bond leads to a loss in binding affinity of between 2 and 6 kJ/mol. H-bonds where one partner is charged exhibit a stronger effect. The mutation Tyr 169 → Phe lowers the binding energy by 15.6 kJ/mol.

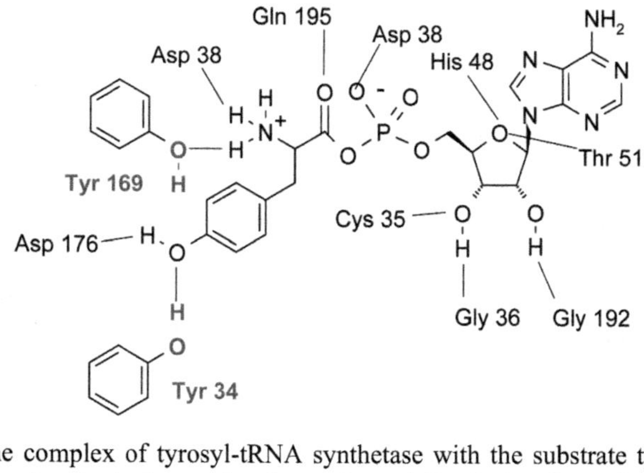

Figure 6. In the complex of tyrosyl-tRNA synthetase with the substrate tyrosyl-adenylate, numerous intermolecular hydrogen bonds are formed. Replacing the amino acid Tyr 34 for Phe or Tyr 169 for Phe means that in each case a hydrogen bond can no longer be formed. This results in a loss of binding affinity.

Fidarestat **1** is a potent aldose reductase (AR) inhibitor. Using its carboxamide group, it forms a hydrogen bond to the NH function of the amide group of Leu 300 (Fig. 7). If the leucine is replaced by proline, the possibility of forming the H-bond is lost as proline does not possess an available NH group. This replacement means a loss of Gibbs free energy of 7.8 kJ/mol. If the partitioning into enthalpy ΔH and entropy $-T\Delta S$ is determined by microcalorimetry, it can be seen that this loss of an H-bond is largely enthalpic in nature (Fig. 7). In comparison, the inhibitor sorbinil **2**, in which the carboxamide group is missing, should be considered. Interestingly, the free binding energy for wild type and Leu300Pro mutant are virtually identical. In sorbinil the group for the formation of the H-bond with the NH group of Leu 300 is missing, accordingly the loss of the NH function in the protein is hardly noticeable. This accounts for the nearly unchanged free binding energy. Nevertheless, the sorbinil complex of the wild type and the mutant are different. The binding to the wild type is enthalpically more favourable, entropically however more expensive compared to the mutant. The crystal structure shows a water molecule pick-up upon sorbinil binding that mediates an H-bond between sorbinil's ether group and the NH function of Leu 300 (Fig. 7). This results in an enthalpic increase of approx. 5 kJ/mol. At the same time, the capturing of a water molecule is entropically unfavourable. This contribution of approx. –6 kJ/mol just compensates the enthalpic gain, so that in total virtually no gain in affinity ΔG remains. The proline mutant cannot form any water mediating contact to sorbinil because of the missing NH function. Hence, the enthalpic gain due to H-bond formation cannot be experienced. However, also no entropic loss as a result of capturing of a water molecule is observed.

The three-dimensional structures of a large number of protein-ligand complexes have been determined. Many of these complexes form hydrogen bonds between protein and ligand. The whole problem concerning individual contributions of the hydrogen bond to binding affinity is summarized in Fig. 8. For 80 protein-ligand complexes, the experimentally determined binding constants are plotted against the number of formed hydrogen bonds. For a given number of hydrogen bonds, the measured binding constants extend over a considerable range. The contribution of an H-bond is therefore by no means constant, but varies obviously. Due to desolvation effects the contribution of an H-bond can even lower the binding affinity. If two ligands are compared that only differ in the functional group that forms the H-bond to the protein, then the affinity may increase, remain constant or even decrease.

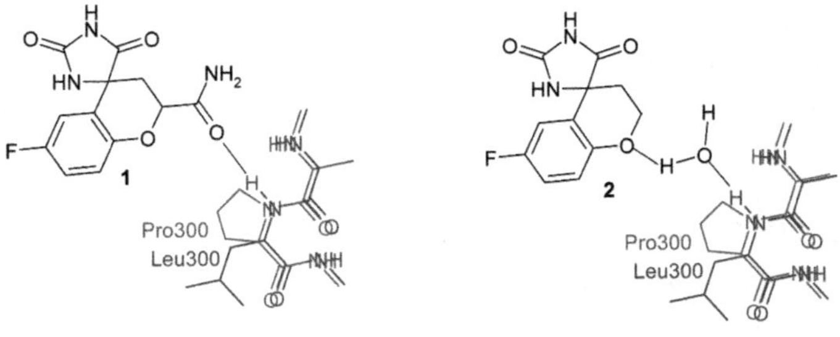

ΔΔG:	7.8 kJ/mol	ΔΔG:	-0.8 kJ/mol
ΔΔH:	6.9 kJ/mol	ΔΔH:	5.1 kJ/mol
-TΔΔS:	0.9 kJ/mol	-TΔΔS:	-5.9 kJ/mol

Figure 7. Fidarestat 1 (left), with its carboxamide group, forms a hydrogen bond to the NH function of Leu 300 (blue). By exchanging the leucine for proline (red), the H-bond can no longer be formed. This leads to a loss in ΔΔG of 7.8 kJ/mol, for which an enthalpic price (ΔΔH: 6.9 kJ/mol) must be paid. Sorbinil 2 (right) lacks a similar carboxamide group. The exchange leucine → proline leaves the Gibbs free binding energy ΔΔG virtually unchanged. However, the binding of sorbinil to the wild type (leucine, blue) is enthalpically more favourable and entropically less favourable than to the proline mutant (red). An interstitial water molecule is picked up by the wild type and mediates an H-bond between sorbinil and Leu 300. This gives the wild type an enthalpy advantage of approx. 5 kJ/mol. At the same time, capturing of a water molecule is entropically unfavourable for the wild type (-TΔΔS: approx. –6 kJ/mol) and compensates the enthalpic advantage.

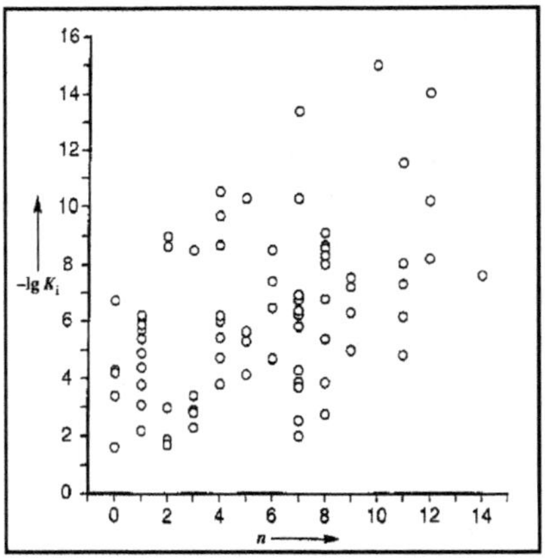

Figure 8. A plot of the binding constants K_i of 80 crystallographically investigated protein-ligand complexes shows that K_i is not a direct function of the number of hydrogen bonds formed between the protein and the ligand.

An impressive example for the importance of hydrogen bonds are the inhibitors **3** of the metalloprotease thermolysin as studied by Paul Bartlett's research group. They replaced phosphonamide -PO_2NH- by phosphinate -PO_2CH_2- and phosphonate -PO_2O-, respectively. The results of this exchange are summarised in Table 2. Although the X-ray structure shows that the NH group forms an H-bond, it can nevertheless be replaced without loss of binding affinity by a CH_2 group. This observation becomes obvious when we, analogous to Fig. 4, compare the inventory of hydrogen bonds before and after ligand binding for the phosphonamide and phosphinate. In both cases the number of H-bonds remains unchanged. If, on the other hand, the NH group is replaced by oxygen, the binding affinity drops by a factor of 1000. In water, the oxygen atom replacing the NH group can form a hydrogen bond to the bulk water. In the phosphonate $-PO_2O-$ complex, however, the electronegative oxygen atom is placed exactly opposite to the oxygen atom of the carbonyl group of Ala 113. Two acceptor groups face each other. A hydrogen bond cannot be formed. The inventory of H-bonds remains unbalanced. Even more, both groups will repel each other resulting in a weaker binding. A similar case can be seen in Table 3. This compares the binding affinities of three thrombin inhibitors **4** synthesised by Eli Lilly. The amine (X = NH) can form an H-bond to Gly 216 and binds most strongly. The ether (X = -O-) binds 5,000 times worse because of the electrostatic repulsion between the ether's oxygen and the protein's carbonyl group. The aliphatic compound (X = $-CH_2-$) exhibits considerable binding that, compared to X = NH, is merely reduced by a factor of 8 (thrombin) and 2 (trypsin), respectively.

TABLE 2. Binding constants Ki for thermolysin inhibitors **3**, with either a phosphonamide (X = -NH-), phosphonate (X = -O-) or phosphinate (X = -CH2-) group. The phosphonamid group -PO2NH- coordinates to the zinc ion simultaneously forming an H-bond to Ala 113.

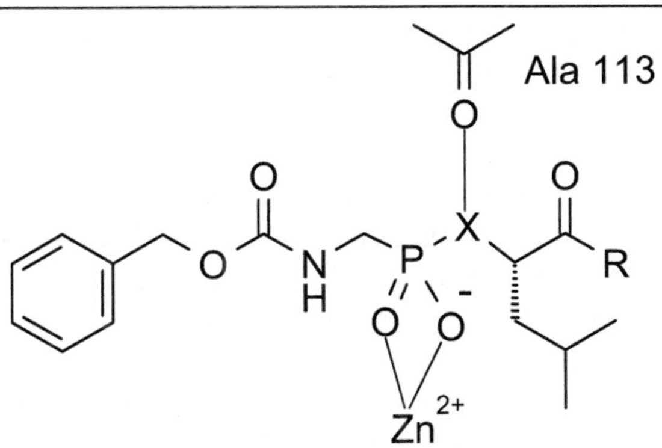

3

R	Binding constants Ki in mM		
	X=–NH-	–O–	–CH2–
OH	0.76	660	1.4
Gly–OH	0.27	230	0.3
Phe–OH	0.08	53	0.07
Ala–OH	0.02	13	0.02
Leu–OH	0.01	9	0.01

TABLE 3. Binding of **4** to the serine proteases thrombin and trypsin.

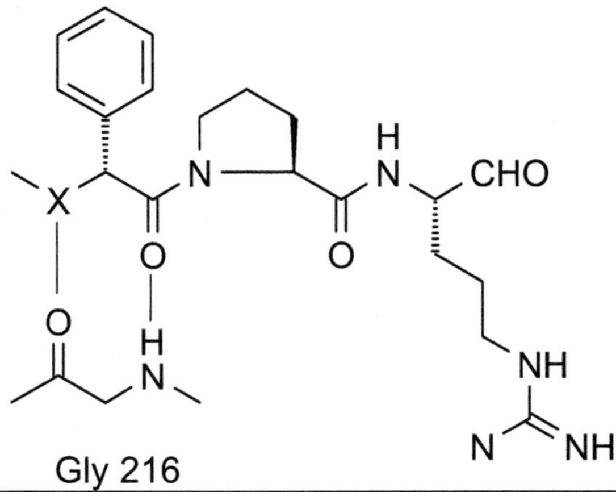

4

Enzyme	IC_{50}-values (mg/mL)		
	X=–NH–	–O–	–CH2–
Thrombin	0.009	52	0.07
Trypsin	0.009	43	0.018

7. The strength of hydrophobic protein–ligand interactions

We have seen that the directly attracting forces between lipophilic groups are significantly smaller than those between polar groups. Hydrophobic interactions are mainly based on the displacement of water molecules. Many experiments have shown that the contribution to the binding affinity is, to a first approximation, directly proportional to the lipophilic surface area that is buried upon ligand binding and is thus no longer accessible to water. Typically a contribution of approximately –50 to –200 J/mol per $Å^2$ lipophilic contact area is found. A case in point is retinol. It binds to the retinol binding protein exclusively by lipophilic contacts with a binding constant of 190 nM. This corresponds to a Gibbs free binding energy of -39.8 kJ/mol, and a lipohilic area of 250 $Å^2$ is buried. The contribution per $Å^2$ for this example amounts to –39,800/250 = –159 J/mol.

Six inhibitors of HIV protease are shown in Fig. 9. In lead optimization **5** was enlarged by attaching hydrophobic groups to its hydrophobic surface. Crystallographically it could be confirmed that the binding mode is thereby not altered. If within the series the changes in molecular volume are plotted against affinity, a linear correlation becomes apparent. Here the binding affinity increases by –65 J/mol$Å^2$.

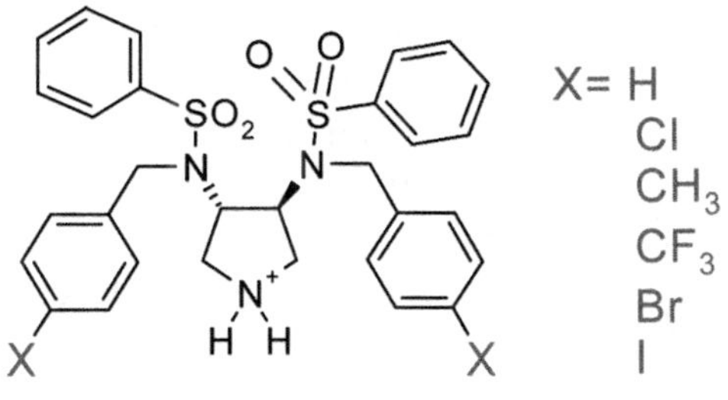

Figure 9. The central skeleton of the HIV protease inhibitor 5 was enlarged in due course of lead optimisation by appending hydrophobic groups in the para-position of the N-benzyl group thus increasing the ligand's surface. It was possible to prove crystallographically that the binding mode remained unchanged. The increasing molecular volume improved binding affinity linearly by –65 J/mol per Å^2.

The hydrophobic interactions are, in many cases, the dominant contribution to the Gibbs free binding energy. Fig. 10 shows, for the same 80 protein-ligand complexes as in Fig. 8, a plot of the lipophilic surface contributions that are buried upon complex formation against the experimentally determined binding constants. Also here a scattering of the values over a wide range is noticed.

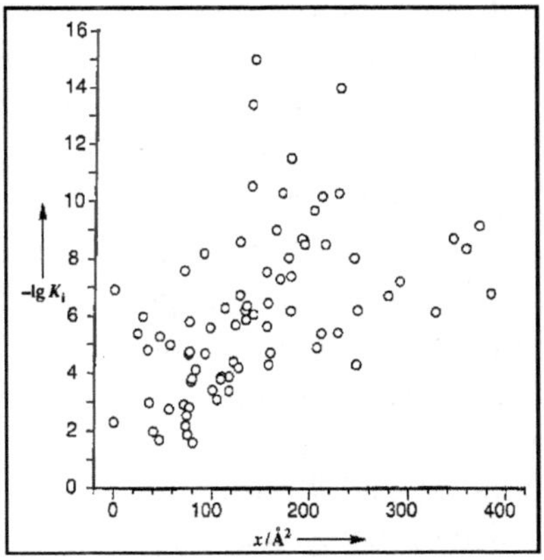

Figure 10. A plot of the binding constants K_i of 80 crystallographically investigated protein-ligand complexes against the surface area buried upon binding. Analogous to Fig. 8, it is indicated that K_i is also not a simple function of the buried surface portion.

8. Binding and mobility: compensation of enthalpy and entropy

Enthalpy and entropy are, according to Eq. 3, closely connected and their sum results in the Gibbs free binding energy. If the formation of protein–ligand complexes is compared, then *ΔG* falls between weakly binding millimolar complexes and strongly binding nanomolar examples in a window of approx. 35–40 kJ/mol. Lead optimisation covers even a smaller range. The binding constant is typically improved by 5–6 orders of magnitude which corresponds to approx. 25–30 kJ/mol. Usually the enthalpy ΔH varies upon functional group replacement in a lead compound over a significantly larger range. If the change in ΔG upon this replacement turns out to be much smaller, the change in enthalpy ΔH must, simply for numeric reasons, be compensated for by opposite changes in entropy $-T\Delta S$. This is the only way how large variations in enthalpy and entropy reveal small changes in ΔG. This provokes an important question: Is there any reason why enthalpy and entropy are opponents and compensate each other, at least in part, during an optimisation? How can it be achieved that both factors are optimised without mutually cancelling out and leaving ΔG virtually unchanged?

Entropy-driven optimisation intends expansion of the hydrophobic surface that is buried upon binding. This very intuitive factor expresses the fact that enlarged ligands displace an increasing number of water molecules upon binding. Usually, the correct freezing of conformational degrees of freedom leads likewise to an improvement to the entropic binding contributions. In order to enhance binding of a ligand to a protein enthalpically, additional polar interactions must be introduced. However, this usually occurs at the expense of the additional polar groups having to first of all lose their water solvation shell. This results in a contribution to desolvation. If, in the thrombin inhibitor **6**, an amidino group is attached to the unsubstituted phenyl group in the *para* position, then for **7** a pronounced improvement of affinity is achieved accompanied by a marked increase in enthalpy (Fig. 11). With the benzamidine group, the inhibitor forms a salt bridge to an aspartate present in thrombin. It is thereby strongly immobilised, which is entropically unfavourable. The inhibitor **6**, which is lacking the polar group, binds with similar geometry. It cannot, however, form the salt bridge. The structure exhibits an increased residual mobility of the inhibitor in the binding pocket, which, from the entropic point of view, is beneficial.

Both of the compounds **8** and **9** represent likewise thrombin inhibitors. They only differ in the size of the cycloalkyl group that has been attached to the basic skeleton in order to fill up a hydrophobic pocket of the protein. Both inhibitors possess the same binding affinity for thrombin. However, their Gibbs free binding energy partitions very differently into enthalpy and entropy. The compound with the cyclopentyl substituent possesses an enthalpic advantage and an entropic disadvantage in comparison to the cyclohexyl derivative.

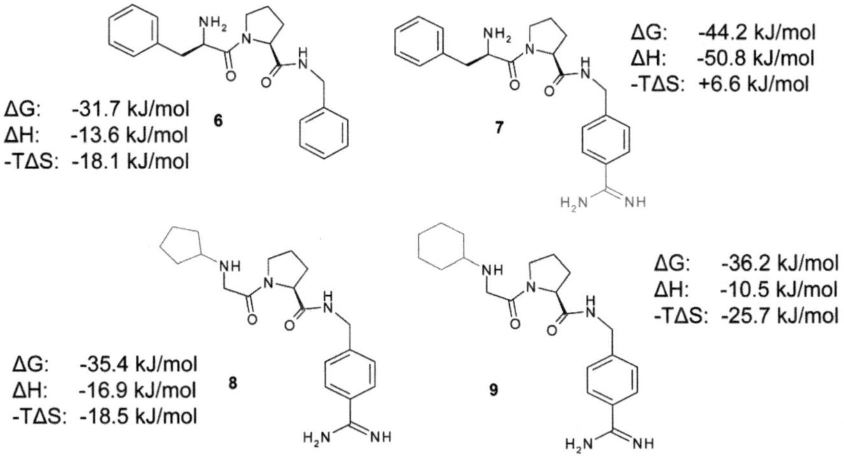

Figure 11. Substitution of the phenyl group in 6 by a para-benzamidinophenyl group in 7 leads, for this thrombin inhibitor, to a pronounced improvement in affinity largely based on an enthalpic gain. This is caused by the formation of a salt bridge to Asp 189. The homologous ligands 8 and 9 bind equally strongly to thrombin, however the binding affinities partitions quite differently into enthalpic and entropic contributions. 9 possesses a significantly higher residual mobility in the binding pocket than 8, thereby providing an entropic advantage for this derivative. But, because of the on average less efficient contacts to the protein, an enthalpic disadvantage is experienced.

Where does this surprising effect originate from? The crystal structures of both derivatives with thrombin show an important difference with respect to the cycloalkyl group. Whereas the five-membered ring can easily be identified in the electron density, practically no density can be observed in the area where the six-membered ring ought to be encountered. Such observations in a crystal structure indicate an enhanced disorder of a particular molecular portion in a protein-ligand complex. This disorder can be of static nature whereby the six-membered ring is distributed over many states. Alternatively, it can possess a much higher residual mobility in the protein-bound state compared to cyclopentyl derivative. Molecular dynamics simulations confirm this difference. In case of the five-membered ring compound, the cyclopentyl group remains in the hydrophobic pocket and performs from time to time a so-called "jump rotation". The virtually planar ring oscillates between two states and interchanges its upper and lower face. This hardly changes the placement of the ring in the pocket. No hydrogen bond is formed by **8** to the carbonyl group of Gly 216. The cyclohexyl ring derivative **9** behaves differently. Here the cycloalkyl group moves in the course of the simulation out of the binding pocket and, after a while, returns back there again. At the same time **9** forms intermediately a hydrogen bond to Gly 216. As a result, **9** possesses in the bound state a high residual mobility.

This difference in the dynamic behaviour between **8** and **9** explains their deviating thermodynamic profiles. The cyclopentyl derivative has an entropic disadvantage as it is more strongly arrested in the binding pocket. The unambiguous orientation provides, however, advantages to form enthalpic interactions. The firm contacts to the protein allows for increased contributions to the interaction energy. Things look different for the six-membered ring derivative. Its reduced immobilization in the binding pocket means a smaller loss in the degrees of freedom during complex formation. This accounts for the entropic advantage, but is enthalpically a disadvantage. During the intermediate release from the binding pocket interactions with the protein can only take place with reduced strength.

What is to be learned from this example? Even if ligands have very similar chemical structures, their binding behaviour can be markedly different. Their residual mobility in the binding pocket can be determinant for the thermodynamic binding contributions. A **mutual compensation between enthalpy and entropy** obviously leads to a nearly unchanged free enthalpy. This interplay of residual mobility in the binding pocket and ability of developing interactions has consequences, of course, for the optimisation process. Medicinal chemists love to think in terms of group contributions experienced during the exchange of certain functional groups to binding affinity. Statistical analyses of such group contributions have been performed and can be used as a set of rules to guide optimisation strategies. This is usually done assuming additivity. How much is gained once a certain group is combined with another one in a molecule upon optimization? However, caution is needed during such considerations. Minor changes in the binding properties cause such simple rules to break down.

The optimisation of the thrombin inhibitor **10** to **11** will be regarded as an example (Fig. 12). Two changes will be made. On the one hand the hydrophobic substituent at one end of the molecular skeleton may be changed from an *n*-propyl to phenethyl group. This signifies a marked increase in the hydrophobic surface area. In a second optimisation step, an amino group is introduced adjacent to the hydrophobic group to form a hydrogen bond to Gly 216. Both these changes from **10** to **11** lead to an improvement in affinity of $\Delta\Delta G = -18.5$ kJ/mol. Both modifications can, of course, be introduced successively via the intermediate steps **12** and **13** respectively. If first the hydrophobic group is enlarged from **10** to **12**, then only a small gain in binding affinity results. If **12** is now further optimised by the introduction of the amino group to the final product **11**, then a distinct gain in affinity is achieved. Does the amino group yield such a strong affinity gain? One can proceed reversely and introduce the amino group first at **10** to reveal **13**. This change accounts for an improvement of merely $\Delta\Delta G = -9.5$kJ/mol. The subsequent enlargement of the hydrophobic surface area from **13** to **11** results in an additional gain in affinity of –9.0 kJ/mol.

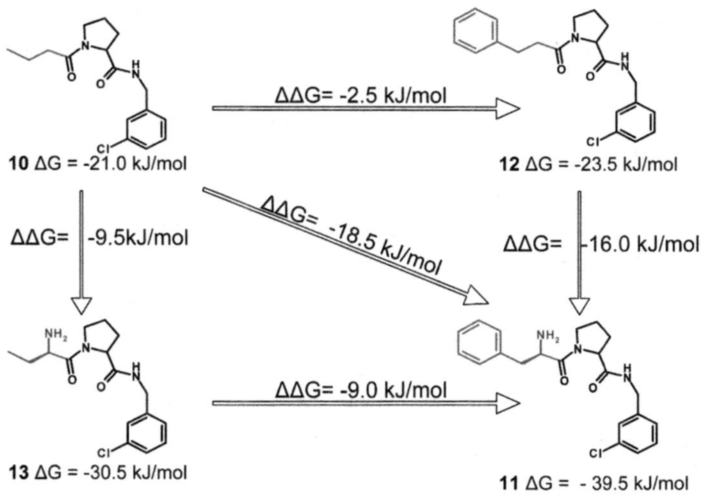

Figure 12. The optimisation of the thrombin inhibitor 10 to 11 means an increase in affinity of $\Delta\Delta G = -18.5$ kJ/mol. The hydrophobic side chains (red) are thereby enlarged from an n-propyl- to phenethyl group and an amino group (blue) is introduced. These changes may also be carried out in a step-wise fashion. Enlarging the hydrophobic surface of 12 improves the affinity by only –2.5 kJ/mol, the major contribution of –16 kJ/mol is provided by the subsequently added amino group. Introducing the amino group first to 13 yields –9.5 kJ/mol, and the subsequent substitution of the hydrophobic residue raises the affinity by a further –9 kJ/mol. The explanation for the non-existent additivity is to be found in the complex interplay of residual mobility, desolvation and the stability of the established enthalpic interactions.

This example shows that simple additivity rules break down. As in the example with the five- and six-membered ring derivatives **8** and **9**, the total balance of the residual mobility, partial solvation of the binding pocket and the strength of the resulting interactions has a determinant influence on the gain of affinity. The interplay of the partially compensating enthalpic and entropic binding contributions is responsible for this complex picture.

9. Lessons for drug design

This chapter should not give the impression that a quantitative prediction of the strength of a protein-ligand interaction is impossible. Despite the complex character of protein-ligand interactions, one ought first to apply some simple rules.

- Many strong protein-ligand interactions are characterized by extensive lipophilic contacts. An extension of the lipophilic contact surface between protein and ligand often leads to an improvement in the binding affinity. This means that the search for unoccupied lipophilic pockets in the protein should be one of the first steps to designing and optimising novel ligands.

- An increase in the binding affinity by means of additional H-bonds is not obvious. An H-bond only contributes to the total inventory if, in the protein-ligand complex, a stronger interaction of the participating groups takes place than in water. On the other hand, the burial of the polar atoms without them being saturated by an H-bond almost always leads to a loss of binding affinity. For the final design of ligands it must be ensured that the polar atoms find partners if, in the protein–ligand complex, they cannot access water.
- Every ligand displaces water molecules when binding to a protein. There are protein binding pockets that are shaped in a way avoiding optimal solvation by water. In such cases a ligand can be in a position to form more H-bonds to the protein than the water molecule can. The binding affinity of such ligands can be very high.
- Rigid ligands can bind more firmly than flexible ones because the loss of internal degrees of freedom is less for rigid ligands.
- Water can form strong H-bonds, but it is often not a comparably potent ligand for transition metals as thiols, acids, hydroxamates and a few other groups are. Accordingly, a direct interaction with the metal ion is important for most proteins that contain an ion of a transition metal. In general, all direct protein-ligand interactions contribute strongly to affinity that either cannot or are only difficult to be replaced by water molecules.

General literature

P.R. Andrews, Drug-Receptor Interactions, in H. Kubinyi, Hrsg., 3D-QSAR in Drug Design. Theory, Methods and Applications, Escom, Leiden, 1993, S. 13–40

P.R. Andrews, D.J. Craik, J.L. Martin, Functional Group Contributions to Drug-Receptor Interactions, *J. Med. Chem.*, **27**, 1648–1657 (1984)

H.J. Böhm und G. Klebe, What Can We Learn From Molecular Recognition in Protein-Ligand Complexes for the Design of New Drugs? *Angew. Chem., Int. Ed. Engl.* **35**, 2588–2614 (1996)T. E. Creighton, Proteins: Structures and Molecular properties, 2nd Ed., W.H. Freeman, New York, 1992

H. Gohlke und G. Klebe, Approaches to the Description and Prediction of Binding Affinity of Small-Molecule Ligands to Macromolecular Receptors, *Angew. Chem., Int. Ed. Engl.* **41**, 2644–2676 (2002)

I.D. Kuntz, K. Chen, K.A. Sharp, P.A. Kollman, The Maximal Affinity of Ligands, *Proc. Natl. Acad. Sci.* USA , 96, 9997–10002 (1999)

Special literature

Ehrlich, P. Chemotherapeutics: Scientific Principles, Methods and Results. Lancet 182, 445–451 (1913)

A. R. Fersht, J. P. Shi, J. Knill-Jones, et al., Hydrogen Binding and Biological Specificity Analysed by Protein Engineering, *Nature* 314, 235–238 (1985)

C. Gerlach, M. Smolinski, et al., Thermodynamic Inhibition Profile of a Cyclopentyl- and a Cyclohexyl Derivative Towards Thrombin: The Same, but for Deviating Reasons, *Angew. Chem. Int. Ed.*, **46**, 8511–8514 (2007)

F. W. Lichtenthaler, 100 Years "Schlüssel-Schloss-Prinzip": What Made Emil Fischer Use this Analogy? *Angew. Chem., Int. Ed. Engl.*, 33, 2364–2374 (1994)

R. P. Mason, D. G. Rhodes und L. G. Herbette, Reevaluating Equilibrium and Kinetic Binding Parameters for Lipophilic Drugs Based on a Structural Model for Drug Interaction with Biological Membranes, *J. Med. Chem.* **34**, 869–877 (1991)

B. P. Morgan, J. M. Scholtz, M. D. Ballinger, I. D. Zipkin und P. A. Bartlett, Differential Binding Energy: A Detailed Evaluation of the Influence of Hydrogen-Binding and Hydrophobic Groups on the Inhibition of Thermolysin by Phosphorous-Containing Inhibitors, *J. Am. Chem. Soc.* **113**, 297–307 (1991)

T. Petrova, H. Steuber, et al., Factorizing Selectivity Determinants of Inhibitor Binding toward Aldose and Aldehyde Reductases: Structural and Thermodynamic Properties of the Aldose Reductase Mutant Leu300Pro-Fidarestat Complex, *J. Med. Chem.*, **48**, 5659–5665 (2005)

STRUCTURE-BASED DESIGN OF TRNA-GUANINE TRANSGLYCOSYLASE INHIBITORS

GERHARD KLEBE*
Institute of Pharmaceutical Chemistry, University of Marburg, Marbacher Weg 6, D35032 Marburg, Germany

Abstract. Taking the development of inhibitors for the tRNA-modifying enzyme tRNA-guanine transglycosylase (TGT) as an example, the scope of a structure-based drug development project will be demonstrated, performed via several cycles of iterative design. The described example is based on studies, performed at ETH-Zurich and University of Marburg in joint collaboration. As these studies have been executed in an academic environment, different tools of structure-based design have been applied and several issues of more fundamental interest to the methodological background of the project could be addressed.

Keywords: Shigellosis, crystal structure, de novo design synthesis, docking, hydrophobic pockets, water network, charge-assisted hydrogen bond

1. Shigellosis - clinical picture and therapeutic approach

Shigellosis is a severe disease caused by bacteria belonging to the genus *Shigella*. An occurring symptom of shigellosis is diarrhea. The bacteria are ingested through contaminated drinking water or food and affect the epithelial cells of the intestinal mucosa. Being extremely infectious, already 10–100 germs may result in an infection. On global level shigellosis represents a serious problem. Every year more than 170 million of diseased persons are recorded, more than 1 million of these infections have lethal consequences. Although the disease occurs primarily in developing countries, there are also reports on more than 1.5 million infected inhabitants of industrial nations.

* To whom correspondence should be addressed. Gerhard Klebe, Institute of Pharmaceutical Chemistry, University of Marburg, Marbacher Weg 6, D35032 Marburg, Germany, fax +49 6421 282 8994; e-mail: klebe@mailer.uni-marburg.de

J.L. Sussman and P. Spadon (eds.), *From Molecules to Medicines*,

Especially under several conditions like missing hygiene, turmoil of war, undersupply of potable water, natural disasters or famines in refugee camps the illness can rage easily. If the onset already happened, shigellosis affects in a first instance children. In many African countries shigellosis reaches a new relevance in combination with AIDS infections.

Like any infectious disease shigellosis is usually first treated with antibiotics. Many of the occurred diseases can be stopped that way in industrial nations. Unfortunately, *Shigella* bacteria, which bear analogy to *Escherichia coli*, intend to create a resistance against antibiotics rapidly. Furthermore, the antibiotics which are supposed to kill *Shigella* bacteria can also affect natural and essential bacteria of the human intestinal flora. On this account many patients, who are treated with antibiotics experience diarrhea and a serious ullage. Especially for infants this can lead to a life-threatening dysfunction of water balance. For all these reasons latest research concentrates on finding a particular therapeutic approach to prevent *Shigella* bacteria from developing pathogenicity.

2. Surpressing pathogenicity development on a molecular level

Shigella bacteria afflict the endothelial cells inside the intestine. To access the human cells they produce own virulence factors called invasines. These are proteins able to form a notable apparatus in combination with proteins of the attacked endothelial cells allowing reproduction of *Shigella* bacteria in infected cells. The required genetic information for virulence factors is stored on plasmids. The actual expression in case of an infection is induced by several transcription factors. Particularly, the factor *VirF* is responsible for the bacteria's development of pathogenicity. To be used efficiently inside the ribosomes, *VirF* relies on specifically modified tRNA molecules. tRNA is a ribonucleic acid consisting of approximately 80 nucleotides. At its end, an tRNA is loaded with an amino acid which corresponds to its specific base triplet in the central loop, the so-called wobble position. While translating the genetic information of the mRNA every encoded amino acid being stored in terms of a base triplet becomes bounded to the ribosome via a correspondent tRNA. This tRNA conveys the right amino acid, so that the correct residue is integrated into the emerging polypeptide chain of the nascent protein. The modifications of the required tRNA involve the nucleobase at position 34 of the wobble position. A chemically modified base must be present. If these modifications are missing, the translation is inefficient. As a consequence the *Shigella* bacteria are no longer able to produce the required amount of invasines to afflict the endothelial cells. For this reason their pathogenicity is perceptibly reduced.

The bacteria possess enzymes which catalyze the required modifications of the nucleobase in tRNA. As a first step, guanine **1** in position 34 is cleaved from the tRNA molecule and replaced by the modified base preQ$_1$ **2** (Fig. 1). In this reaction the enzyme tRNA-guanine transglycosylase (TGT) serves as a catalyst. Subsequently, the exchanged base in tRNA is further modified by an enzymatic cascade until the product base queuine is produced. According to these named effects, inhibitors of TGT represent a specific therapeutic principle enabling selective intervention into the development of pathogenicity by *Shigella* bacteria. In contrast to a therapy based on broad-spectrum antibiotics the bacteria are not killed but become detained from affecting the endothelial cells pathogenically.

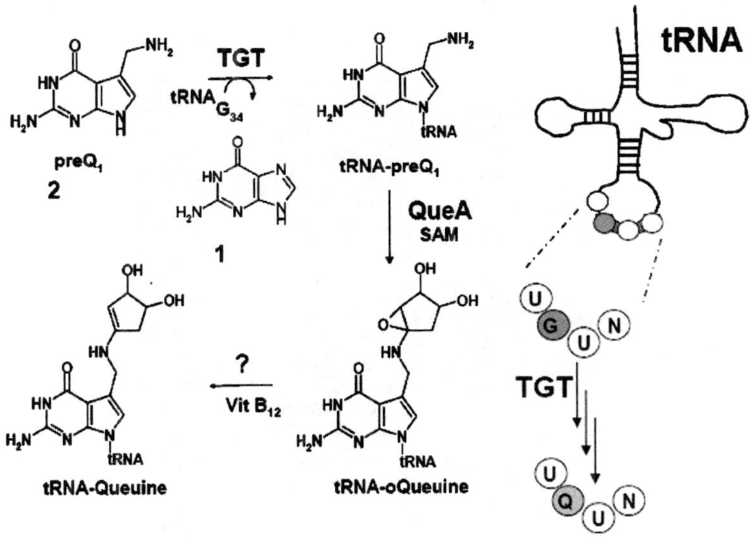

Figure 1. The enzyme tRNA-guanine transglycosylase (TGT) catalyzes the replacement of guanine 1 by preQ1 2 in tRNA (left). Subsequently, further modification of this base are performed by different enzymes to reveal as final product queuine in tRNA. The replacement of the base occurs at the wobble position of tRNA (right).

3. Starting point: the crystal structure of tRNA-guanine transglycosylase

In the first instance, the determination of the crystal structure of TGT in complex with preQ$_1$ was successfully performed with a related species. In the active site of the latter enzyme only Phe is exchanged by Tyr, an exchance of no influence to ligand binding. At a later stage, also the structure of TGT could be determined in a complex with a truncated tRNA (Fig. 2). Accordingly to this structure, the base exchange (Fig. 3) proceeds as indicated in the following reaction scheme. At first the tRNA is bound with the covalently attached base guanine. The base is pulled out of the tRNA molecule together with the ribose sugar. It is specifically recognized by Asp 102, Asp 156, Gln 203 and Gly 203. Asp 280 attacks as a nucleophile the carbon atom C1 of the ribose ring. The C1-N bond is cleaved and guanine is released. The base leaves together with a water molecule the binding pocket and $preQ_1$ is acommodated at the same site. To achieve this task, the peptide bond between Leu 231 and Ala 232 has to flip. Subsequent to the release of a proton, the basic nitrogen of $preQ_1$ attacks nucleophilically the ribose sugar, which is covalently attached to Asp 280. Once the new bond to the tRNA is formed, the modified tRNA leaves the enzyme. Asp 102 is essential for the recognition process of the bound base. In addition, most likely this amino acid provides the required protons in due course of the mechanism or accepts them in a subsequent step.

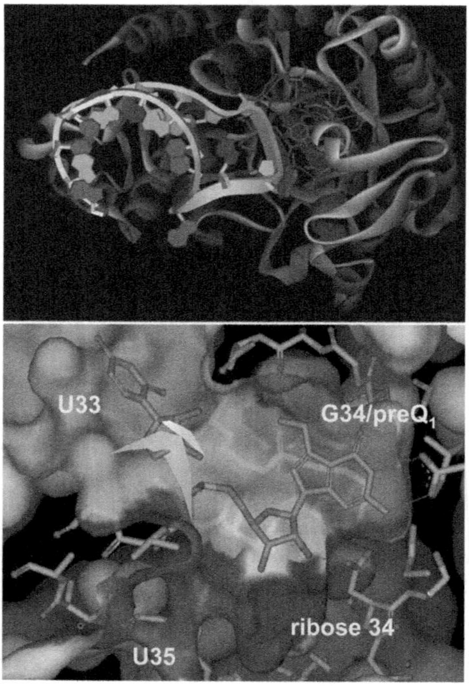

Figure 2. The crystal structure of TGT with a portion of bound tRNA. The protein adopts TIM-barrel fold. The tRNA binds with its bases U33, G34 and U35 to the protein. Thereby, the base to be exchanged is completely pulled out of the tRNA-molecule in position 34. On the right hand side, a view into the binding pocket is shown. The incorporated modified base $preQ_1$ is held in position in the guanine recognition pocket (orange) by Asp 156, Asp 102, Gly 230 and Leu 231. The ribose sugar occupies to a small hydrophobic pocket (blue). Uracil 33, which is located in the sequence prior to the exchanged base is places into the binding pocket indicated in green, while the subsequent uracil 35 is found in the binding area highlighted in red.

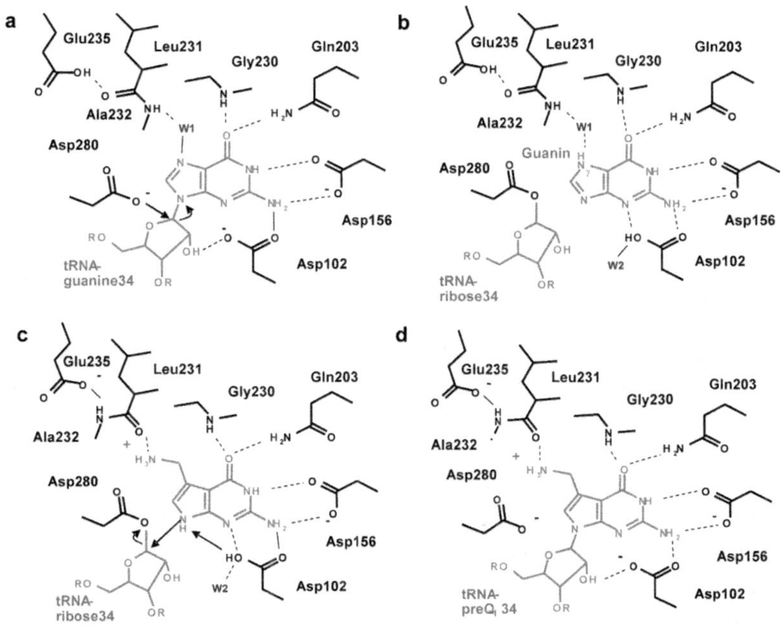

Figure 3. Mechanism of base exchange in the glycosidase. tRNA is bound with guanine 34 and a water molecule mediating contact to nitrogen in 3-position. Asp 280 attacks nucleophilically carbon C1of the ribose ring. The C1-N bond is cleaved and guanine is released. It leaves together with a water molecule the binding pocket. Into the same binding site $preQ_1$ is accommodated, the peptide bond between Leu 231 and Ala 232 has to flip simultaneously. After deprotonation, the basic nitrogen of $preQ_1$ starts nucleophilic attack onto the ribose sugar which is covalently bound to Asp 280. The newly formed modified tRNA is released from the enzyme.

4. A functional assay for the determination of binding constants

The exchange reaction of the base occurs in two steps. In principle, both steps can be blocked by inhibitors. This fact has to be considered in a functional assay. In a first step, the unmodified tRNA is bound (Fig. 4). Sufficiently large inhibitors might prevent this step competitively. Once the tRNA is covalently attached to the enzyme, the base guanine is released and leaves the protein. As a next step, $preQ_1$ is bound. During this reaction step, however, also a potential inhibitor might compete for the occupation of the specific $preQ_1$ binding site. This competing inhibitor may not be much larger than guanine or $preQ_1$. Therefore, especially smaller inhibitors show a different inhibition profile compared to the larger ones.

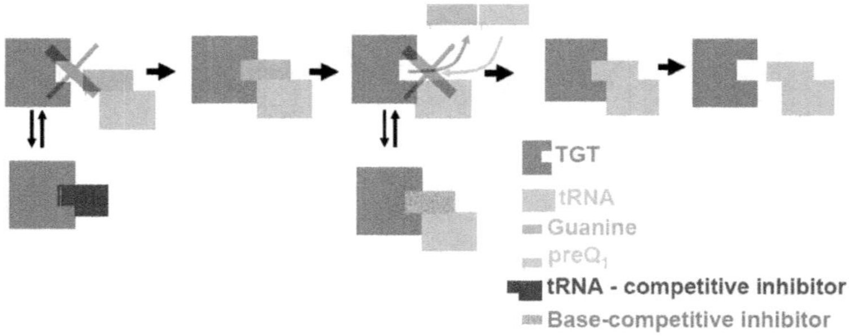

Figure 4. The base exchange reaction proceeds via two steps. Inhibitors can either compete with binding of the complete tRNA (left, dark grey) or with the substitution of the small nucleobase (centre, light grey).

To record inhibition radioactively labeled guanine is used. If this guanine reacts with tRNA, TGT catalyzes the incorporation of the radioactively labeled base. In consequence tRNA becomes radioactively labeled. Precipitating the tRNA from the reaction conditions in fixed time intervals and measuring the integrated radioactivity, the reaction kinetics of the incorporation process and, at the same time, also the catalytic rate of the enzyme can be recorded. If potential inhibitors are added, there are less TGT molecules available for implementation and the incorporation rate declines. This becomes noticeable in the observed enzyme kinetics. Being evaluated in detail the enzyme kinetics allow determination of inhibition constants. Additionally, it is also possible to differentiate, whether the inhibitors interact competitively with the entire tRNA or whether they just compete with the exchange of the small base.

5. LUDI discovers first lead compounds

At the beginning of the research project only the structure of the binary complex of TGT with $preQ_1$ was available. Furthermore, the two-stage inhibition mechanism, explained in the last paragraph, was not yet characterized. At a later stage, it was Bernhard Stengl who succeeded to elucidate the details of this process in due course of the described project. Ulrich Grädler took the binary TGT·$preQ_1$ complex as a reference and applying the de novo design program LUDI he could embark into the search for putative inhibitors. In a chemical catalogue he could detect some first hits. The ligands listed in Fig. 5 had been suggested. Compound **3** turned out to be a micromolar inhibitor. Together with this hit a crystal structure could be determined (Fig. 6). To great pleasure, the newly discovered lead 4-aminophthalhydrazide **3** bound in exactly the same fashion as predicted by LUDI. In the next step, LUDI was consulted to suggest substituents for the novel skeleton that could fill unoccupied space in the binding pocket. At first, expansion of the ring system by an additional aromatic ring was proposed. Furthermore, placement of a nitrogen-containing heterocycle onto unsatisfied interaction sites close to Asp 102 and Asp 280 were taken in consideration. Hans-Dieter Gerber succeeded to synthesize the derivatives **4–6** (Fig. 7). **4** and **5** revealed tenfold better inhibition in the enzymatic assay. The more disappointing was the fact, that the heterocyclic derivate **6** exhibited less potent binding compared to the initial lead compound **3**. Ulrich Grädler succeeded to resolve the crystal structures with this inhibitor. It showed the expected binding mode. In the complex with **6** the heterocycle was found in short contact distance to the terminal amide group of Asn 70. Hence, it appeared obvious to attach an additional amino group to **6**, in order to form a further contact to the protein. The synthesis was successful and the crystal structure of **7** showed the expected binding mode with an additional H-bond. However, also this derivative dropped in affinity compared to the original parent lead structure **3**. Detailed analysis of the structural data indicated for either **6** and **7**, a disorder of the attached heterocyclic moiety and a strongly expanded, rather weak hydrogen bond between the exocyclic amino group and the carbonyl group of Leu 231 of the peptide backbone. Incorporation of the heterocycle at a position next to the neighboring aspartate residues followed the idea that this moiety, potentially present in positively charged state and additionally capable to involve in H-bond formation, would be ideal. Presumably, both protein residues adopt a deprotonated state. In this case, a positive charge at the interstitial triazole moiety would be best suited to interact with its neighbourhood. But which charged state can be assumed for the heterocycle? For a related model compound pK_a measurements have been performed. Furthermore, under the same buffer conditions as applied to grow the crystals for the protein structure determination, crystals of the small molecule ligand alone could be obtained. Both measurements revealed that the heterocycle is present in uncharged state with both neighboring nitrogens in deprotonated state. Even though not necessarily the same protonation state must be given at the protein-binding site, the following model appeared conclusive to explain the reduced affinity of **6** and **7**. At the interstitial position between both negatively charged aspartate residues the uncharged triazole moiety experiences at least with one of the two neighboring acidic protein residues repulsive interactions. This would explain the dropping binding affinity, the observed disorder and the expanded H-bond distance to the carbonyl group of Leu 231.

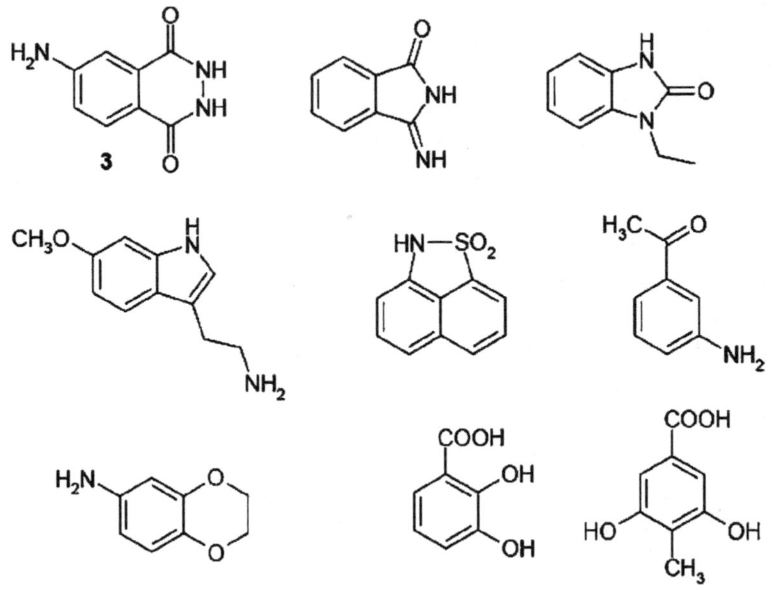

Figure 5. Suggestions for first leads obtained by LUDI. Among these three was tested and turned out to be a two-digit micromolar inhibitor.

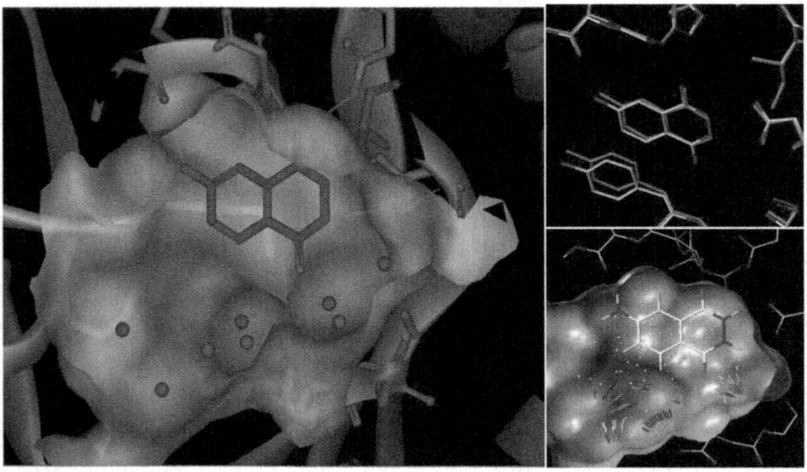

Figure 6. Crystal structure of TGT with 3, the first hit from LUDI. The agreement of predicted and experimentally confirmed binding modes is virtually perfect (upper right). LUDI suggests in the lower part of the binding pocket still unsatisfied interaction sites.

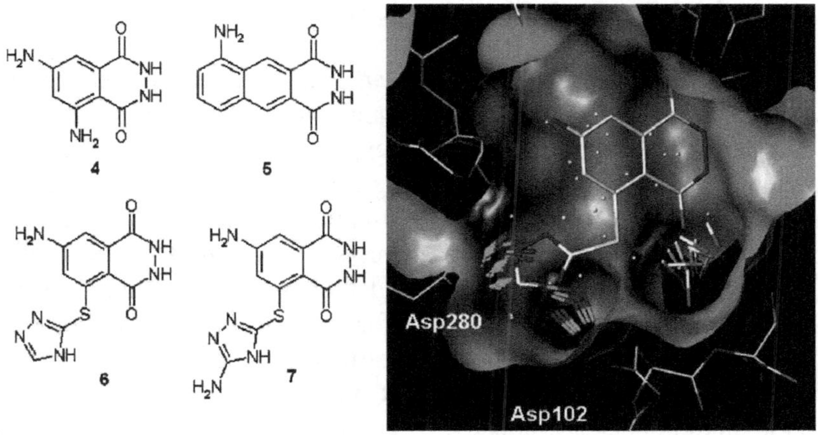

Figure 7. Based on 3, 4 and 5 were developed to occupy the unsatisfied interaction sites, indicated in Fig. 6. They turned out to be tenfold better binders. To further exploit the unused additional interaction sites heterocycles (6 and 7) were added to the basic skeleton. Both derivates show reduced binding affinity, probably caused by repulsive interactions with the two neighboring aspartate residues 102 and 280. The heterocycles do not bear the desired favorable positive partial charge.

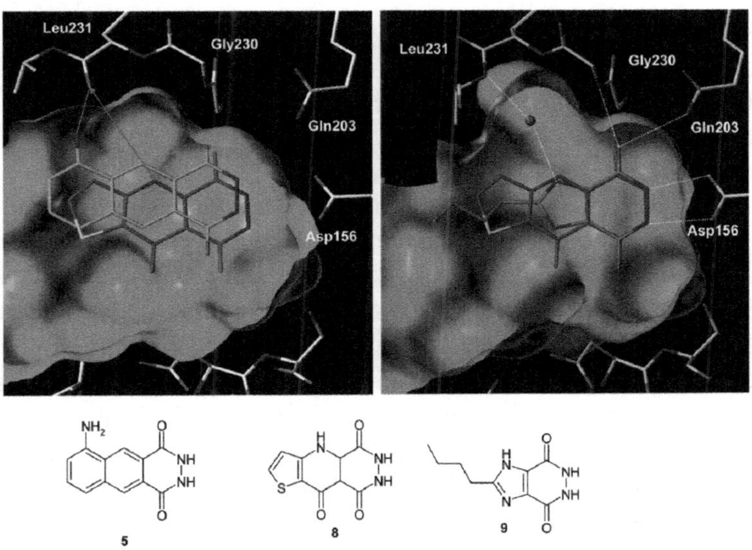

Figure 8. Also 8 was supposed to agree with the interaction pattern indicated by the initial lead. Docking studies place this derivative into the binding pocket (violet, left) with a distance much too large for a hydrogen bond between the polar nitrogen in the central pyridazinone ring and the carbonyl group at Leu 231. Nevertheless, 8 binds micromolar to the protein. The crystal structure (right) determined with a similar inhibitor 9 (orange) shows two surprising results: The peptide bond flips its orientation and exposes its NH-group towards the binding pocket. In addition, a water molecule (red sphere) mediates an interaction of the ligand to the protein!

6. Surprise, surprise: a flipped amide bond and a water molecule

By courtesy Novo Nordisk provided **8**, another compound which perpetuates a very similar interaction pattern as the initial lead. Nevertheless, the docking of this derivative indicated a strongly expanded distance between the polar nitrogen in the central pyridazinone ring and the carbonyl group of Leu 231, much too large for a hydrogen bond formation. Nevertheless, the compound turned out to be a micromolar hit. The crystal structure determined with the help of the very similar derivate **9** provides an explanation (Fig. 8). A peptide bond, occurring in two conformations for mechanistical reasons, operates as a type of functional switch. In the new structure it flips over to adopt reverse orientation! With flipped geometry, it now presents its NH functionality into the binding pocket. Mediated by an interstitial water molecule the contact between this NH group and the polar nitrogen of the ligand is established. At the current state of the project, the above-described enzymatic mechanism had not been elucidated to all details, thus the flip of the peptide bond was a real surprise and could not be predicted. Furthermore, the pick-up of a water molecule presented a second big surprise. This fact underlines the importance of determining crystal structures concurrently for any newly discovered lead structures.

7. Hot-spot analysis and virtual screening opens the floodgate to new ideas for synthesis

How might the observation of multiple binding modes be made a virtue? Ruth Brenk used both the protein conformers observed in the complex with **3** and **9** to perform a hot-spot analysis. The result of these evaluations is shown in Fig. 9. The pharmacophore derived from this analysis was used as starting point for virtual screening which discovered a plenty of alternative molecules to accommodate the guanine binding site (Fig. 10). Almost all of the discovered and tested hits proved to serve as micromolar inhibitors. They provided a whole bunch of new ideas for the synthesis of alternative inhibitors. Three skeletons were selected for the follow-up studies. Their structures deduce from a pyridazinone (triones, **10**), pteridine (**11**) or 6-amino-quinazolinone moiety (**12**), respectively (Fig. 11). The latter skeleton can be imagined as being assembled of the right-hand part of the natural substrate guanine **1** and the left-hand part of the first successful hit **3** discovered by LUDI.

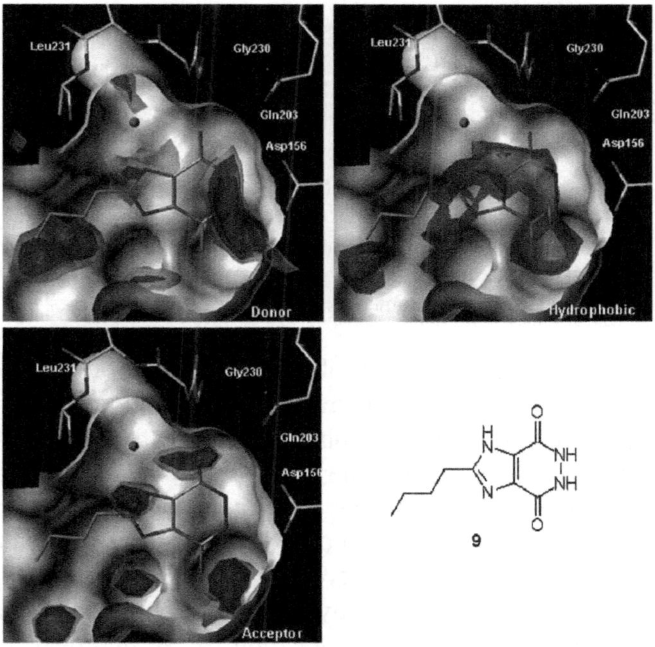

Figure 9. Hot-spot analysis for preferred binding sites for a hydrogen-bond donor (top), acceptor (centre) and hydrophobic functional groups (low). Additionally, 9 is shown, which occupies favored binding sites with its polar groups. In the lower left part of the binding pocket (close to the binding site of the ribose sugar, Fig 1, blue) additional binding sites are indicated, which have been addressed in the following design step.

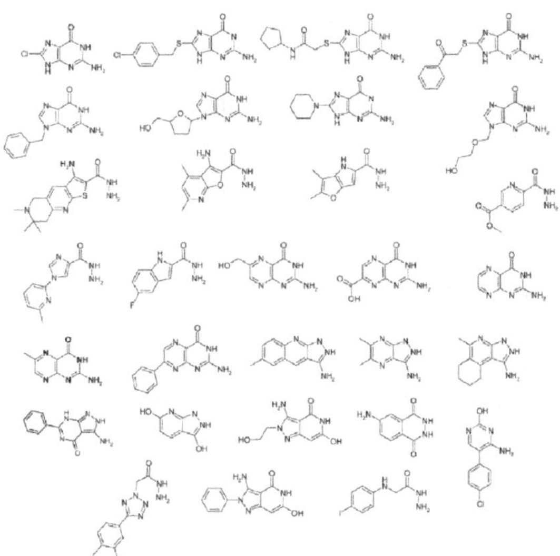

Figure 10. Hits from a virtual screening campaign. Out of these, several examples exhibited mircromolar inhibition.

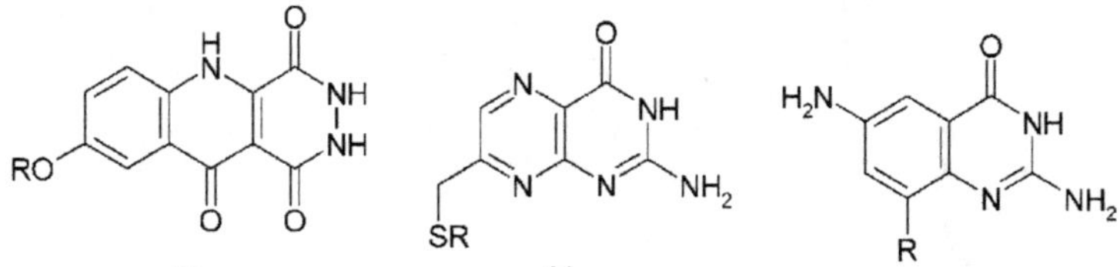

10 **11** **12**

Figure 11. The pyridazinone (trione, 10), pteridine (11) and 6-amino-quinazolinone scaffolds (12) were considered as putative leads for further synthesis and lead optimization. Via attachment of suitable side chains R numerous derivatives could be synthesized.

Let us return to the distribution of hot spots in the binding pocket. The new lead compounds all occupy interaction sites found in the upper part of the binding pocket. However, some other areas "down left" next to aspartate 102 and 280 either indicate placement of potential H-bond donor sites or hydrophobic building blocks. So far these areas were not exploited in the design. Referring to the binding mode of bound tRNA (Fig. 1), this area accommodates the ribose sugar moiety at position 34. The hot-spot analysis suggests occupation of this area by a hydrophobic molecular portion. Slightly above this area a favorable location for an hydrogen bond donor functionality is suggested. This region conforms to the binding site between both aspartates which has already been exploited as a putative interaction site in the design of the heterocyclic derivatives **6** and **7**. Laterally to this pocket, another favorable region for an H-bond acceptor group is suggested. Here, in the natural substrate tRNA, the ribose sugar moiety at position 34 with its 2'- or 3'-hydroxyl groups is placed.

8. About the filling of hydrophobic pockets and interference with a water network

A golden rule in drug design suggests that filling of unoccupied hydrophobic pockets with lipophilic side chains, attached to a lead skeleton, will improve binding affinity. Inhibitors with appropriate side chains were designed and resulted in the derivates listed in Fig. 12. Disappointingly, after synthesis they showed just an insignificant improvement. Besides the pteridines and aminoquinolinones in Zurich also lin-benzoguanine **13** were picked up by synthesis as a potential lead skeleton. Joint efforts of Emanuel Meyer and Simone Hörner at ETH Zurich soon provided a whole series of novel inhibitors, that allowed for detailed crystallographic analyses and establishment of structure-activity relationships. Surprisingly, for none of the derivatives, listed in Fig. 12, a really significant improvement of binding affinity could be observed. They occupy, as initially planned, a small hydrophobic pocket created next to Val 45, Leu 68 and Asn 70. Mutual comparison of the individual crystal structures indicated, that the protein had to perform pronounced induced-fit adaptations upon accommodation of these inhibitors. However, this is also observed for the natural substrate (Fig. 3) and the adaptations are very similar to those, observed upon tRNA binding. Accordingly, it appears rather unlikely that these adaptations and the opening of the pocket will be very costly in energy. Otherwise, the enzyme would experience difficulties in binding its own substrate sufficiently. Therefore, a too high barrier in energy seemed unlikely as a straight-forward explanation for the lacking affinity improvement of the designed inhibitors. Bernhard Stengl and Tina Ritschel analyzed the different derivatives again with great detail. Interestingly enough, the small parent skeletons achieved already one-digit micromolar inhibition. The attachment of additional small substituents, which orient in the direction of the hydrophobic pocket, resulted in an affinity loss. Only the filling of the small hydrophobic pocket with the attached aromatic side chains could compensate for this initial loss of affinity and recover one-digit micromolar binding. Further insights revealed a comparison of the water molecules present in this area of the binding pocket in the different inhibitor crystal structures. In the small unsubstituted derivatives a cluster of water molecules form a network of hydrogen bonds between both, most likely charged aspartate residues 102 and 280. This water network provides an essential part to the solvation of the two polar acidic residues. Possibly, it also serves as a kind of buffer to stabilize the strong charges in this part of the binding pocket. All derivatives, displayed in Fig. 12, bridge this area, filled in the uncomplexed situation by the water network, with a hydrophobic chain to place the terminal hydrophobic substituent into the small hydrophobic pocket. In consequence, they interfere with the water network which is paid for by a loss in affinity (Fig. 13)!

Remarkable is an affinity comparison of compounds **15** and **16** (Fig. 14). The derivative **15** with the 7-dimethylamino group at the quinazolinone moiety drops in binding affinity by a factor of 10 compared with the unsubstituted derivative. Replacement of one of the methyl groups by a benzyl portion in **16**, partially recovers potency. The crystal structure with this derivative shows that the benzyl group is not placed towards the small hydrophobic pocket but orients in direction to the pocket that recognizes uracil 33 in the natural substrate (Fig. 2). A new design concept became apparent: No hydrophobic chains should penetrate and perturb the water network mediating between Asp 102 and Asp 280. Instead, a hydrophobic moiety should be attached to the parent skeleton of the ligands which allows for placement into the uracil 33 recognition pocket.

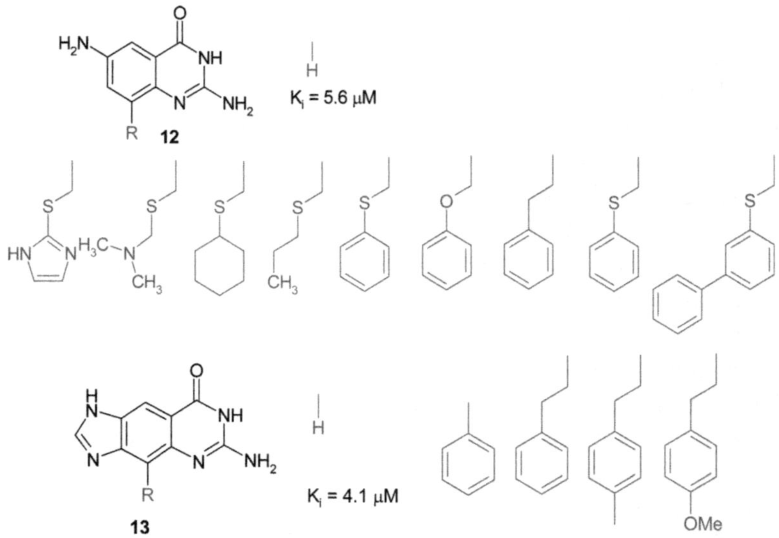

Figure 12. The 6-amino-quinazolinone scaffolds 12 could be further derivatized by adding different substituents R. Surprisingly, even the best compounds from the series did not exceed beyond one-digit micromolar binding. As an alternative scaffold lin-benzoguanine 13 has been picked up and substituted in 4-position. Despite very good inhibition of the parent structure, all substituted derivatives could not achieve significantly improved binding affinity.

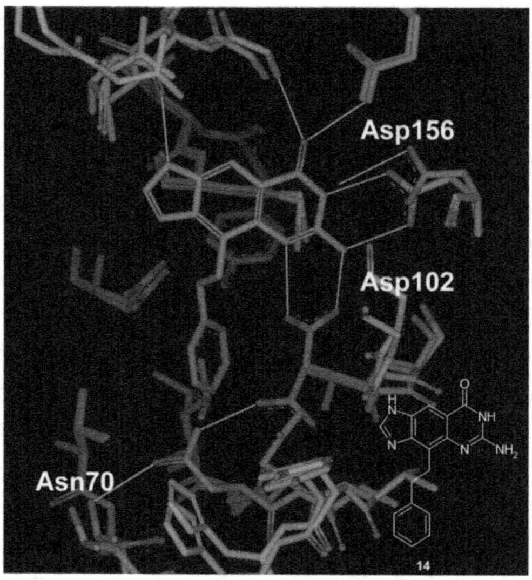

Figure 13. A comparison of the crystal structure of the uncomplexed protein with the bound complex of 14 demonstrates pronounced induced-fit adaptations of the protein upon ligand binding. They also occur once the natural substrate tRNA is accommodated. They result in an opening of a small hydrophobic pocket flanked by Val 45, Leu 68 and Asn 70.

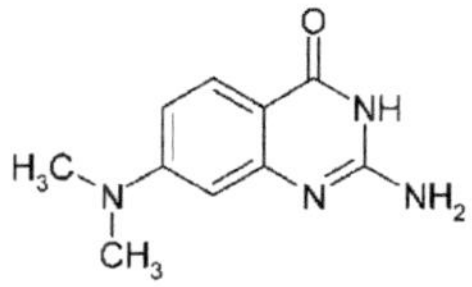

15 K_i = 31±10μM

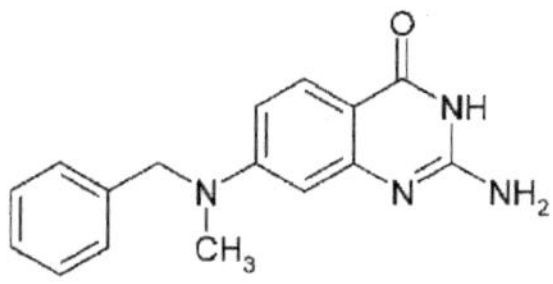

16 K_i = 7.6±3.7μM

Figure 14. The quinazolinone derivative 15 with a 7-dimethylamino group loses binding affinity by a factor of 10 compared to the unsubstituted parent structure. Replacement of one of the two methyl groups by a benzyl substituent (16) recovers part of the binding affinity. The crystal structure of the benzyl derivative shows that the benzyl moiety is not oriented into the small hydrophobic pocket but towards the uracil 33 pocket.

9. With a salt bridge: finally nanomolar!

Synthetically the desired side chain orientation could be best achieved using the lin-benzoguanine skeleton as parent structure. The unsubstituted lin-benzoguanine **13** shows a water network involving five distinct water molecules in the crystal structure (Fig. 15). Attachment of a methyl group (**17**) in 2-position, improves binding by a factor of 2.7. Replacement of this methyl group by a 2-amino group (**18**) further improves the binding constant dramatically by a factor of 20. Introduction of the amino group in 2-position at the lin-benzoguanine moiety reveals in total an enhancement by a factor of 50. How can this surprising effect be explained? In section 5 the importance of a hydrogen bond to the backbone carbonyl group of Leu 231 has been discussed. This functional group is located in the peptide bond which turned out to operate as a molecular switch in due course of the project. During synthesis of additional inhibitors, we had to learn that this hydrogen bond, assumed on first glance as essential for binding, accounts only in some derivatives for a pronounced contribution to binding affinity (compare **21** with **22** or **23** with **24**, Fig. 16). The lin-benzoguanine moiety also forms this H-bond to the carbonyl group of Leu231. Upon introduction of the 2-amino group, the central imidazole moiety is modified into a guanidinium-type group. Such a change corresponds to a strong shift in basicity of the skeleton. Experimentally determined pK_a-values confirm this jump in pK_a by more than one order of magnitude. Calculations to model this pK_a-shift upon complex formation indicate an even further shift towards basic properties. In consequence, the compounds should bind to the protein in protonated state, thus bearing a positive charge in this region of the imidazole portion. As a result, the original hydrogen bond is modified into a charge-assisted hydrogen bond. As such, it exhibits a much stronger contribution to binding affinity and explains the experienced strong increase in potency.

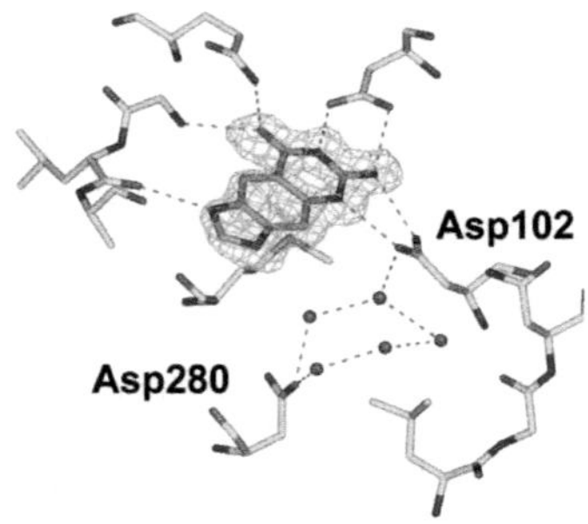

Figure 15. The parent scaffold of lin-benzoguanine 13 binds with $K_i = 4.1$ μM to the protein and leave the water network (red spheres) between the most likely negatively charged Asp 102 and Asp 280 intact.

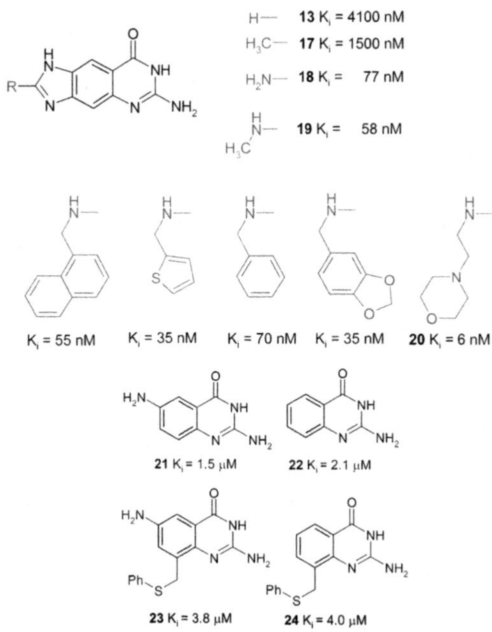

Figure 16. Substitution of the lin-benzoguanine scaffold in 2-position reveals a significant improvement of binding affinity. Particularly, the introduction of a 2-amino group (18) reveals derivatives with strongly shifted basic character so that they will bind in positively charged state. In consequence, they form charge-assisted H-bonds with the carbonyl group of Leu 231 resulting in a significantly enhanced contribution to affinity. A comparison of 21 and 22 or 23 and 24 underlines, that this hydrogen bond only contributes to binding potency if supported by the presence of a positive charge in this molecular portion. Once this charge is lacking, replacement of this H-bond forming amino group does not affect binding affinity. The morpholino derivative 20 displays a nanomolar inhibitor and orients its side chain into the uracil 33 pocket.

As already indicated, the filling of the uracil 33 binding pocket parallels a further increase in affinity. Accordingly, different substituents were attached to the 2-amino group. A methylene linker has been introduced, however, in order to decouple attached aromatic portions from an electronic conjugation with the 2-amino group. Among the synthesized derivatives the morpholino derivative **20** turned out to be the most strongest binder with a K_i= 6nM. It also exhibits best water solubility. Interestingly enough, in all crystallographically studied derivatives of this series the attached side chains cannot be localized in the difference electron density. Most likely, their side chains are disordered over multiple states. In principle this argues against strong enthalpic binding of these groups in the pocket. However, for entropic reasons this detrimental effect can be compensated and in total a significant improvement of the Gibbs free energy of binding is experienced.

The development of nanomolar inhibitors for TGT had to pass through several ups and downs. Determinant for the final success were the multiple crystal structure analyses of protein-ligand complexes. As a take-home message regarding drug design and optimization, the following three aspects can be learnt. The perturbation with an essential water network can be quite detrimental to affinity gain. The filling of a hydrophobic pocket is most likely affinity enhancing, however, it has to be checked whether the linker used to drive the placement of the hydrophobic groups in the requested pocket actually provides an ideal interaction pattern. The exchange of a neutral by a charge-assisted hydrogen bond can take significant influence on binding during the lead optimization process. This can be achieved by adding molecular portions that provoke a significant shift of the pk_a properties of the ligands.

References

R. Brenk, L. Naerum, U. Grädler, H.D. Gerber, G.A. Garcia, K. Reuter, M.T. Stubbs and G. Klebe. Virtual Screening for Submicromolar Leads of TGT based on a New Unexpected Binding Mode Detected by Crystal Structure Analysis, *J. Med. Chem.*, **46**, 1133–1143(2003)

U. Grädler, H.-D. Gerber, D.A.M. Goodenough-Lashua, G.A. Garcia, R. Ficner, K. Reuter, M.T. Stubbs and G. Klebe. A New Target for Shigellosis: Rational Design and Crystallographic Studies of Inhibitors of tRNA-Guanine Transglycosylase, *J. Mol. Biol.*, **306**, 455–467(2001)

S. Hörtner, T. Ritschel, B. Stengl, Ch. Kramer, G. Klebe, F. Diederich. Design, Synthesis, and Biological Evaluation of Inhibitors of tRNA-Guanine Transglycosylase, an Enzyme linked to the Pathogenicity of the Shigella Bacterium. *Angew. Chem. Int. Ed.* **46**, 8266–8269(2007).

E.A. Meyer, R. Brenk, R.K. Castellano, M. Furler, G. Klebe, F. Diederich. De Novo Design, Synthesis, and in Vitro Evaluation of Inhibitors for Prokaryotic tRNA-Guanine Transglycosylase (TGT): A Dramatic Sulfur Effect on Binding Affinity, *ChemBioChem*, **2**, 250–253(2002)

B. Stengl, K. Reuter and G. Klebe. Mechanism and substrate specificity of tRNA - guanine transglycosylases (TGTs): tRNA modifying enzymes from thee three different kingdoms of life seem to share a common mechanism. *ChemBioChem* **6**, 1–15(2005)

B. Stengl, E.A. Meyer, A. Heine, R. Brenk, F. Diederich and G. Klebe. Crystal structures of tRNA-guanine transglycosylase (TGT) in complex with novel and potent inhibitors unravel pronounced induced-fit adaptations and suggest dimer formation upon substrate binding. *J. Mol. Biol.*, **370**, 492–511(2007)

PROGRESS ON NEW HEPATITIS C VIRUS TARGETS: NS2 AND NS5A

JOSEPH MARCOTRIGIANO*
Department of Chemistry and Chemical Biology,
Center for Advanced Biotechnology and Medicine,
Rutgers University, NJ, USA

Abstract. Hepatitis C virus (HCV) is a major global health problem, affecting about 170 million people worldwide. Chronic infection can lead to cirrhosis and liver cancer. The replication machine of HCV is a multi-subunit membrane associated complex, consisting of nonstructural proteins (NS2-5B), which replicate the viral RNA genome. The structures of NS5A and NS2 were recently determined. NS5A is an essential replicase component that also modulates numerous cellular processes ranging from innate immunity to cell growth and survival. The structure reveals a novel protein fold, a new zinc coordination motif, a disulfide bond and a dimer interface. Analysis of molecular surfaces suggests the location of the membrane interaction surface of NS5A, as well as hypothetical protein and RNA binding sites. NS2 is one of two virally encoded proteases that are required for processing the viral polyprotein into the mature nonstructural proteins. NS2 is a dimeric cysteine protease with two composite active sites. For each active site, the catalytic histidine and glutamate residues are contributed by one monomer and the nucleophilic cysteine by the other. The C-terminal residues remain coordinated in the two active sites, predicting an inactive post-cleavage form. The structure also reveals possible sites of membrane interaction, a rare cis-proline residue, and highly conserved dimer contacts. The novel features of both structures have changed the current view of HCV polyprotein replication and present new opportunities for antiviral drug design.

Keywords: Crystal structure, macromolecular architecture, disulfide bridges, zinc binding site, protein interaction surfaces

* To whom correspondence should be addressed. Joseph Marcotrigiano, Department of Chemistry and Chemical Biology, Center for Advanced Biotechnology and Medicine, Rutgers University, NJ, USA; e-mail: jmarco@cabm.rutgers.edu

J.L. Sussman and P. Spadon (eds.), *From Molecules to Medicines*,

1. Introduction

Hepatitis C virus (HCV) continues to be a major public health problem. In most cases, HCV infection becomes chronic and can persist for decades, leading to cirrhosis, end-stage liver disease and hepatocellular carcinoma. Currently, 3% of the human population – approximately 170 million people – is infected with HCV. In the developed world, HCV infection is the most common cause of liver transplantation and results in 10,000 to 20,000 deaths a year in the US.[5] There is no vaccine, and current HCV therapy, pegylated interferon-alpha in combination with ribavirin, leads to a sustained response in only 50% of genotype 1-infected patients, the prevalent genotype in the US.[11] The current HCV treatment has numerous side effects, causing many patients to prematurely stop treatment. Given the high prevalence of infection and poor response rate, inhibitors that specifically target HCV proteins with fewer side effects are desperately needed. In addition, an effective vaccine would greatly reduce the spread of the virus.

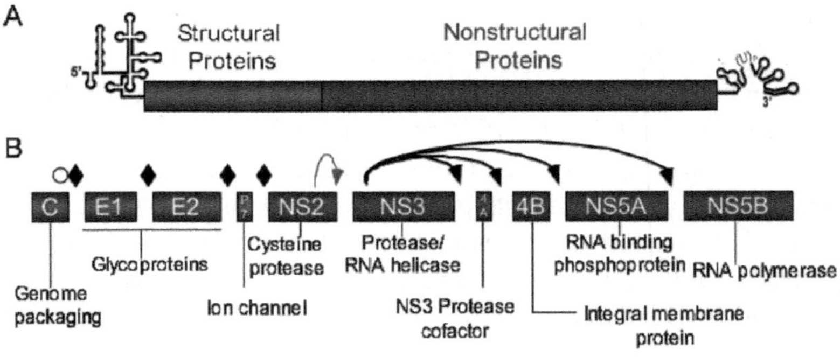

Figure 1. (A) The organization of the HCV genome showing the 5' and 3' NTRs. The open reading frame is represented by the rectangle and is colored grey for the structural proteins and blue for the nonstructural proteins. (B) Polyprotein processing scheme. The black diamonds and open circle denote the cleavage sites for signal peptidase and signal peptide peptidase, respectively. The arrows signify the cleavages performed by the viral encoded NS3-4A (black) and NS2-3 (red). A brief description of each protein is given.

HCV is a member of the family *Flaviviridae*, which also includes Pestiviruses and Flaviviruses. Since its identification in 1989,[22,44] phylogenetic analysis of various isolates has resulted in the classification of six distinct genotypes that are further divided into a number of subtypes (e.g. 1a, 1b, 1c, etc.). The HCV virion consists of an enveloped nucleocapsid containing the viral genome, a single-stranded, positive sense RNA that encodes a single open reading frame[25] (Fig. 1). Once the virus penetrates a permissive cell, the HCV genome is released into the cytosol where the viral RNA is translated in a cap-independent manner by an internal ribosome entry site (IRES) located within the 5' nontranslated region (NTR). Translation generates a viral polyprotein that is proteolytically processed by cellular and viral encoded proteases into ten proteins (Fig. 1). The N-terminal region of the polyprotein is cleaved by cellular signal peptidase and signal peptide peptidase to yield the structural components of the virus particle (core, envelope proteins E1 and E2) and an ion channel (p7). The mature non-structural proteins (NS2, NS3, NS4A, NS4B, NS5A, and NS5B) are liberated by two essential virus-encoded enzymes: the NS2-3 cysteine protease and the NS3-4A serine protease. NS3-5B comprise the minimal viral proteins necessary to form the RNA replication machinery or replicase. HCV replication occurs in association with the perinuclear and ER membranes, utilizing both cellular and viral proteins. Replication involves the synthesis of a genome-length, minus strand that serves as a template for the production of new positive strands for packaging. Not much is known about HCV assembly and egress, since systems to study these processes have not been available until recently. However, extrapolations have been made from comparisons with other flaviviruses. HCV virion assembly is thought to occur on the ER membrane. Newly synthesized, genomic RNAs are encapsulated by core. These nucleocapids bud into the ER, encircling it with the envelope membrane and HCV glycoproteins. The virions travel through the secretory pathway and are released at the cell membrane.

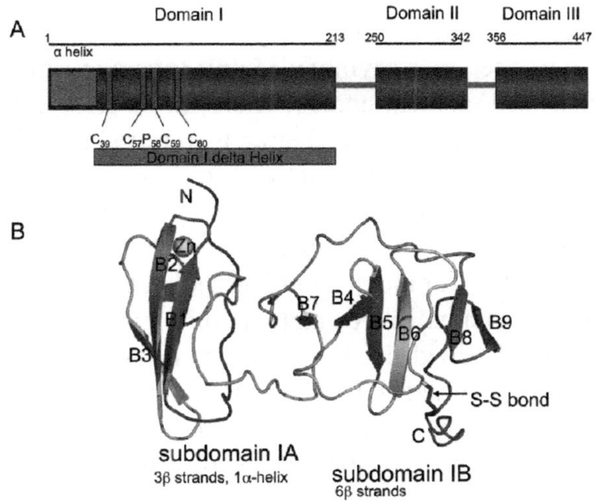

Figure 2. An overview of the NS5A domain I structure. (A) Schematic representation of the domain structure of NS5A. The crystallization construct is denoted by the red bar. (B) Ribbon diagram of the structure of domain I. The polypeptide is colored from the N terminus (blue) to the C terminus (red).

2. HCV NS5A

NS5A is an active component of the HCV RNA replicase, a pivotal regulator of replication, and a modulator of numerous cellular processes spanning from innate immunity to apoptosis and disregulated cell growth.[29,41] It exists as a serine phosphoprotein present in hypo- (56 kDa) and hyper-phosphorylated (58 kDa) forms. An amphipathic α-helix at its N terminus promotes association with cellular membranes.[4,8,32] Following this helix, NS5A is organized into three domains (Fig. 2A).[40] The N-terminal domain (domain I) coordinates a single zinc atom per protein molecule.[40] Mutations disrupting either the membrane anchor[8,32] or zinc binding[40] are lethal for HCV RNA replication. NS5A interacts with other viral components of the replicase[7] and has been shown to modulate NS5B polymerase activity *in vitro*.[37,38] NS5A domain II and the region connecting domain I and II are hotspots for adaptive mutations that can enhance replication in cell culture by more that 10,000-fold.[2,26] Such highly adaptive mutations downregulate NS5A hyperphosphorylation and promote interaction with hVAP-A, a t-SNARE that is required for HCV RNA replication.[10,43,47] In addition to its role in RNA replication, NS5A antagonizes dsRNA activation of PKR[39] and interacts with numerous cellular proteins to modulate cell growth and survival.

3. Architecture of NS5A domain I

The crystal structure of domain I (amino acids 36–198) reveals two essentially identical monomers per asymmetric unit (average r.m.s.d of 0.4 Å) packed together as a dimer via contacts near the N-terminal ends of the molecules. Domain I consists of nine β-strands (referred to as B1–B9) and a single short α-helix connected by a network of extended loops (Fig. 2B). This helix has been designated H2 to allow numbering of the previously determined N-terminal membrane anchoring helix H1.[32] These loop regions comprise a surprising 62% of the domain I structure, with a number of large (14 or more amino acids) extended random coil elements. For ease of discussion, the molecule is divided into two subdomains; the N-terminal subdomain IA containing β-strands B1, B2 and B3, and the α-helix H2, and the C-terminal subdomain IB containing the remaining 6 β-strands (Fig. 2B). The structure of subdomain IA consists of an N-terminal extended loop lying adjacent to a three-stranded anti-parallel β-sheet, with H2 at the C-terminus of the third β-strand. The elements comprise the structural scaffold for a four-cysteine zinc atom coordination site at one end of the β-sheet. A long stretch of random coil exits H2 and passes across the β-sheet and

N-terminal strand in an orientation orthogonal to the plane of the sheet (green in Fig. 2B). This coiled sequence then enters a tight turn and runs adjacent to the β-sheet away from the zinc atom, towards what is referred to as the 'bottom' of subdomain IA, before entering a proline rich region that connects subdomain IA to subdomain IB. Domain IB consists of a four-strand anti-parallel β-sheet (B4, B5, B6, and B7) and a small two-strand anti-parallel β-sheet near the C-terminus (B8 and B9) surrounded by extensive random coil structures. A disulfide bond in subdomain IB connects the B6/B7 loop and the C-terminal loop region. The search for proteins with similar fold architecture to domain I using the DALI[16] server was unsuccessful at identifying related structures, indicating domain I represents a unique protein fold.

4. The NS5A zinc binding site

We previously demonstrated the coordination of a single zinc atom by the domain I region of NS5A.[40] The coordination of this metal ion is absolutely required for RNA replication. The location of the four cysteines involved in zinc coordination (Cys 39, Cys 57, Cys 59, and Cys 80), in relation to predicted secondary structures surrounding these residues, suggested a model of the zinc binding site, that of a four stranded, anti-parallel β-sheet with the zinc atom coordinated at one end of the sheet.[40] Overall, the proposed model is quite similar to the actual organization of the NS5A zinc-binding site revealed by the structure. The four cysteine ligands all lie within loop regions. Cys 39 is within the large N-terminal loop, Cys 57 and Cys 59 are positioned in the loop between strands B1 and B2, and Cys 80 is positioned in the long loop connecting B3 to H2. The distances of the cysteine side chain sulfur groups to the zinc atom for Cys 39 (2.36 Å), Cys 57 (2.47 Å), Cys 59 (2.42 Å) and Cys 80 (2.45 Å) observed in domain I are close to the ideal 2.35 (+/- 0.09) Å distances for structural metal zinc coordination sites in proteins.[1] Similarly, the side chain geometries are within the acceptable limits of previously published values.[1]

DALI[16] database searches of the subdomain IA region were unable to locate similar metal binding folds. A survey of zinc containing proteins with known structures was also unsuccessful at identifying proteins that coordinate metal ions like NS5A. The location of the zinc-binding site in the structure of domain I, combined with previous biochemical characterization, strongly suggests this ion plays a structural role in NS5A fold maintenance. The presence of a novel metal ion coordination motif, combined with the previous demonstration that zinc binding is essential for replication, provides an interesting potential anti-viral drug design target.

5. Potential for a cytoplasmic disulfide bond

Perhaps the most surprising observation from model building and refinement of the structure was the presence of a disulfide bond near the C-terminus of domain I. The disulfide bond connects the sidechains of the conserved Cys 142 and Cys 190, resulting in a covalent link between the loop exiting from β-strand B6 to the C-terminal extension of strand B9 (Fig. 3A). Model refinement without a disulfide at this position placed the sidechains of these cysteine residues in an unfavorable proximity. Refinement with the disulfide led to no problematic geometry for either cysteine residues, and generated a model that better fit the electron density (Fig. 3A). Density corresponding to the disulfide bond is present in both molecules of domain I in the asymmetric unit, providing two independent views of this feature. The disulfide bond in the model results in a sulfur to sulfur atom distance of 2.03Å, an ideal value for bond formation.[9] Analysis of the purified domain I protein in the presence and absence of reducing agent, demonstrated that the oxidized form of the protein runs slightly faster than the reduced form, consistent with the existence of a disulfide bond (Fig. 3B). Therefore, the disulfide is present in the protein and is not an artifact of the crystallization conditions. Although all evidence points to the presence of a disulfide bond in the protein, it is not yet clear if this bond exists in NS5A in the context of an HCV infection.

It has long been held that disulfide bonds only occur in cytoplasmic proteins that are involved in oxidative/reductive chemistry. However, recent publications have shown that the cytoplasm is not as reducing as originally thought, and that transient disulfide bonds can be formed.[30] It is important to consider that NS5A is primarily present not in the cytoplasm at large, but in a replicase complex, a structure involving considerable ER membrane alterations that represents a unique cytoplasmic microenvironment in which an oxidizing environment could exist. A number of thiol containing cytoplasmic proteins have been shown to use reactive oxygen species (ROS) to rapidly regulate protein conformation and activity via reversible disulfide bond formation.[23] These proteins use ROS as a switch to regulate their activities by metal ion release and subsequent disulfide bond formation under conditions of oxidative stress. NS5A has been shown to induce oxidative stress and ROS in the context of replicating viral RNA in cell culture, leading to an inhibition of the anti-apoptotic properties of NF-κB.[12] However, the role of ROS in HCV biology is controversial, as ROS have been shown to both stimulate and inhibit HCV RNA replication in cell culture.[6,33] It is not yet clear what role, if any, is played by ROS and NS5A in HCV biology, but the presence of a potential regulatory disulfide bond in NS5A makes this an interesting and testable hypothesis. Independent of ROS, a number of cytoplasmic proteins, including those of viruses, have been shown to contain transient disulfide bonds, often serving regulatory activities for these proteins.[24,35] Perhaps this is the case for NS5A. Mutagenesis of the Cys 142 and Cys 190 residues produced no measurable defect in HCV RNA replication, indicating the disulfide bond is not required for the replicase functions of NS5A.[40] However, it is enticing to imagine that the disulfide bond, tethering the C-terminus of domain I and likely altering the arrangement of the C-terminal domain II and III, plays a regulatory role in NS5A function, serving as a conformation switch to modulate functions of NS5A.

6. A dimeric NS5A reveals potential molecular interaction surfaces

A surface of domain I that was of considerable interest was the buried region between monomers to create the dimeric domain I seen in the crystal structure (Fig. 4). The dimer interface of 678 Å^2 consists of two patches of buried surface area, one located primarily in subdomain IA, and a second located in subdomain IB. The total buried surface area in the domain I dimer is more than the generally accepted standard of 600 Å^2 for protein interfaces and has good shape and electrostatic complementarity.[19] The contact patch in subdomain IA contains a number of conserved residues, whereas the smaller patch in subdomain IB is of lower sequence conservation. It is important to note that residues of lower conservation involved in the dimer interface appear to primarily be involved in mainchain contacts, suggesting some sequence plasticity is allowable at these positions. The zinc-binding site is close to the interface between monomers and may represent a factor involved in NS5A oligomerization. Homotypic oligomeric interactions of the NS5A protein in yeast two-hybrid and immunoprecipitation experiments have recently been described[7]. Preliminary analytical ultracentrifugation experiments suggest domain I is monomeric in solution, although it is important to note that these experiments were performed with only a portion of NS5A protein in the absence of membranes or nucleic acids (data not shown). Perhaps the conditions that favor protein-protein interactions in crystallization have captured a glimpse of part of the relevant NS5A oligomerization interactions. Clearly more work is needed to determine the oligomeric state of the NS5A protein. Nonetheless, the dimeric form of domain I observed in the crystal structure provides a number of interesting features that may have relevance to HCV biology. Fig. 4A displays a side view of the dimer showing a large groove between the domain IB regions of chain A and B generated by the "claw-like" shape of the dimer.

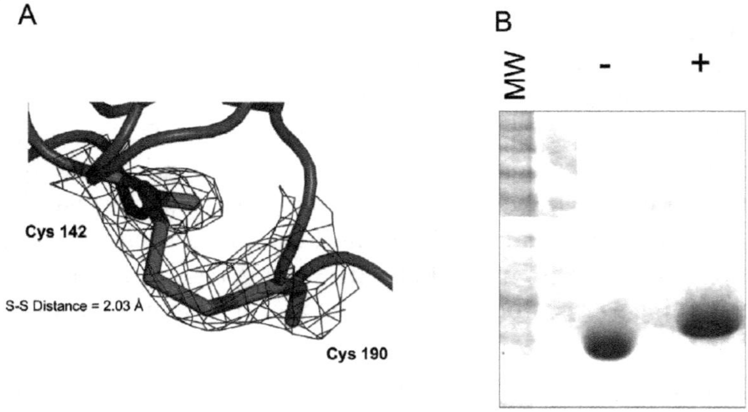

Figure 3. NS5A disulfide bond. (A) Overlay of the experimental electron density map contoured at 1σ on the model of the disulfide bond. Amino acid residues Cys142 and Cys190 are labeled. (B) SDS-PAGE analysis of the purified domain I protein in the presence (+) and absence (-) of 5mM DTT.

As NS5A likely makes contacts with proteins, RNA, and membranes, an analysis of the protein electrostatic surface potential is especially important in determining how a dimeric domain I might fit into the replicase (Fig. 4A). One interesting feature in these surface potential plots is the very basic surface of domain I where the N-termini of the dimer are located (toward the top of the page in Fig. 4A).

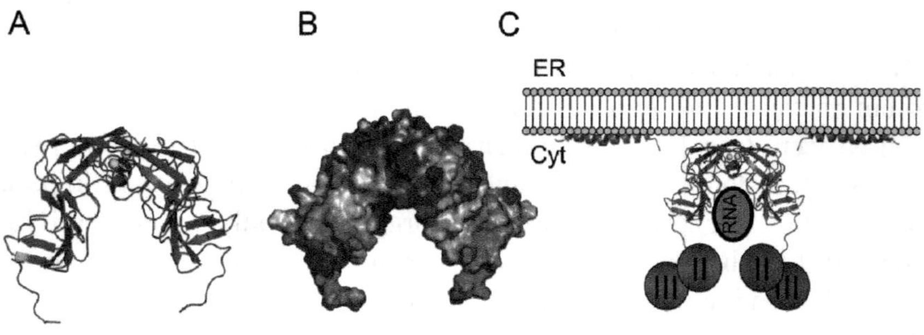

Figure 4. NS5A domain I dimer. Ribbons diagram (A) and the corresponding surface potential (B) of the NS5A crystallographic dimer. (C) Model of the NS5A dimer position relative to the ER membrane, highlighting possible RNA binding surface and NS5A domains II and III.

An NMR structure of the first 31 amino acids of NS5A has recently shown that amino acids 5–25 form the H1 amphipathic helix membrane anchor.[32] The NMR H1 structure and our domain I crystal structure are separated by only five amino acids, suggesting the N-terminus of domain I is likely close to the membrane (Fig. 4B). The presence of a basic surface on the portion of domain I close to the N-terminus is logical, as the protein is likely to be in close contact with the negatively charged lipid head groups of the membrane. This interaction would position the groove generated by the domain I dimer to face away from the membrane where it could interact with RNA. This large groove is an attractive nucleic acid binding pocket, especially when surface potentials of the highly basic groove are plotted. Modeling studies suggest this groove is of sufficient dimensions to bind to either single or double stranded RNA molecules. The deep, highly basic portion (the inside portion of the "claw-like" structure) of the groove has a diameter of 13.5 Å. The more hydrophobic boundary of the basic groove is considerably larger, with a diameter of 27.5 Å. A dsRNA molecule with a diameter of approximately 20 Å, could easily fit in the groove, making both electrostatic contacts with the deep basic groove and hydrophobic contacts with the groove boundary region. The electrostatic character of the groove is conserved among HCV sequences, suggesting this may be an important functional feature. Interactions of NS5A with nucleic acids during protein purification have been described,[17] although specific RNA binding of NS5A remains to be demonstrated. The 'arms' extending out past this groove are more acidic, perhaps serving as a clamp to prevent RNA from exiting the groove. The inside of the acidic arms lining the groove are separated by about 32 Å. The positioning of NS5A domain II and III relative to the groove remains unclear, but it is interesting to imagine these domains interact with the RNA positioned by the domain I groove (Fig. 4B). The relevance of this model to HCV biology by biochemical and genetic characterization of the NS5A protein is currently being evaluated. Nonetheless, the ability to crudely position a dimeric domain I relative to the membrane surface and identify possible RNA and protein interaction surfaces provides a wealth of information to investigate the interactions of NS5A within the replicase.

7. HCV NS2

The nonstructural proteins, NS2-NS5B, are processed by two viral proteases, the NS2-3 autoprotease and the NS3-4A serine protease (Fig. 1). Remarkably, these two viral proteases overlap but function independently. The NS2-3 autoprotease mediates a single cleavage at the NS2/3 junction and encompasses the C-terminal half of NS2 and the N-terminal third of NS313-15,31,36,42. The N-terminal third of NS3 encodes a serine protease catalytic domain that requires the cofactor NS4A and a structural zinc atom to form a fully functional enzyme cleaving at four downstream sites. Although only NS3-NS5B are necessary to form a functional RNA replicase, the activities of both viral proteases are required for HCV replication in cell culture[18] and in chimpanzees.[21]

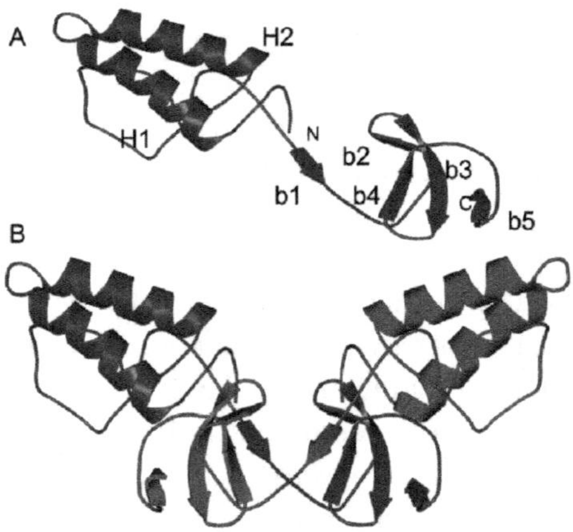

Figure 5. (A) Ribbon diagram showing the NS2pro monomer. The termini and secondary structure elements are labeled. (B) Ribbon diagram of the NS2pro dimer with the two molecules colored red and blue.

The NS3-4A serine protease has been extensively studied and small molecule inhibitors are beginning to enter clinical trials. In contrast, the catalytic mechanism of the NS2-3 autoprotease was controversial. The minimal region required for cleavage at the NS2/3 junction contains no obvious homology to proteolytic enzymes other than the NS3 serine protease domain, which can be inactivated by mutagenesis without affecting NS2-3 cleavage. The observation that NS2-3 cleavage was inhibited by metal chelators and stimulated by addition of zinc led to the speculation that NS2-3 was a metalloprotease.[14] However, the structure of the serine protease domain of NS3 revealed a structural zinc ion chelated by three cysteine residues and a water molecule.[20,28] Thus, limiting zinc could inhibit the NS2-3 autoprotease by affecting the folding of the serine protease domain rather than by playing a direct role in catalysis. Consistent with this idea, mutation of the zinc-coordinating residues in NS3 inhibited autoprotease activity.[13,14] The identification of essential histidine (His 143) and cysteine (Cys 184) residues, and an important glutamate (Glu 163) residue in NS2, led to the alternative proposal of a cysteine protease catalytic mechanism[13,14,46]. We described the structure of the minimal region of NS2 (residues 94 to 217) necessary for autoproteolysis to 2.3 Å resolution.[27] The structure revealed that NS2 is a dimer and contains two cysteine protease active sites with amino acids from both molecules contributing to each catalytic triad.

8. Architecture of the NS2 protease domain

NS2 is a hydrophobic protein with several putative transmembrane segments. Removal of the N-terminal region containing these segments greatly increased the expression and yield of recombinant NS2 without affecting autoproteolysis.[31,42] The C-terminal domain of NS2 residues 94 to 217 ($NS2^{pro}$) was expressed in bacteria and purified to homogeneity in the presence of detergents. The crystal structure revealed that $NS2^{pro}$ consists of two subdomains connected by an extended linker (Fig. 5A). The N-terminal subdomain begins with two antiparallel α helices (H1 and H2) connected by a short loop. Following the second helix, the subdomain has a random-coil conformation that contacts both H1 and H2. The protein then continues into a long, extended coil before entering the antiparallel β sheet in the second subdomain. The last β strand (b5) continues to the C terminus of NS2.

Analysis of the higher order structure within the asymmetric unit revealed that $NS2^{pro}$ is organized into a packed dimer (Fig. 5B). The two molecules of the dimer have a total buried surface area of about 1,300 Å^2 with good electrostatic and shape complementarity.[19] The overall shape of the dimer resembles a 'butterfly', with two-fold symmetry along the vertical axis in Fig. 5B. The N-terminal subdomain of one molecule in the dimer interacts with the C-terminal subdomain of the other molecule and vice versa. Surprisingly, the two molecules contain a long linker that crosses over in the middle of each molecule. The linker forms a β strand (b1) with the antiparallel β sheet in the C-terminal subdomain of the other molecule in the dimer. The N termini of the two monomers lie relatively close to each other, with a distance of 17 Å between their alpha-carbon atoms. The C termini are positioned on opposite sides of the molecule and are solvent-exposed.

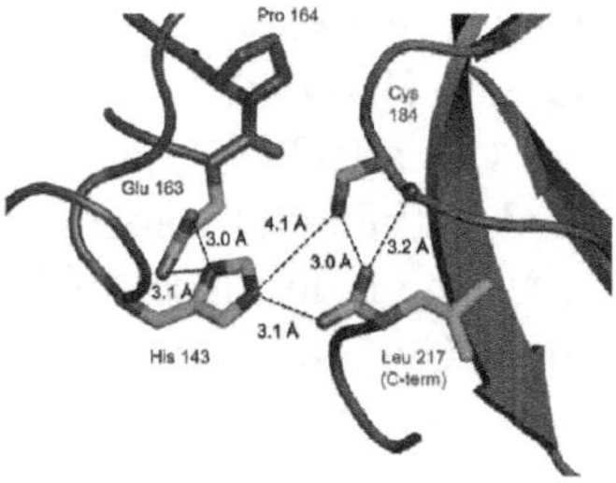

Figure 6. The active site of NS2pro. Residues His 143, Glu 163, Pro 164, Cys 184 and Leu 217 are shown as stick drawings. The active site is composed of His 143 and Glu 163 from one molecule of the dimer (blue) and Cys 184 from the other molecule (red). The C-terminal residue, Leu 217, originates from the red molecule. Dashed lines indicated contacts between atoms and the lengths are provided.

9. NS2 is a cysteine protease

Earlier experiments identified His 143, Glu 163, and Cys 184 as critical residues for NS2-3 proteolytic activity.[13–15] His 143 and Glu 163 are located in the large random coil region following helix H2 in the N-terminal subdomain, while Cys 184 lies at the end of the linker arm in the b1-b2 loop in the C-terminal subdomain. Within the monomer, the side chains of His 143 and Glu 163 form hydrogen bonds, while Cys 184 is separated by approximately 35Å. However, the His and Glu from one monomer lie in close proximity to the Cys from the other chain in a solvent-exposed groove at the dimer interface (Fig. 6). The arrangement of these three residues suggests the formation of a composite cysteine protease active site.

A search of the database of known protein folds using the DALI server[16] with the $NS2^{pro}$ structure found several unrelated proteins with low statistical significance (Z-score less the 3.0). Although $NS2^{pro}$ may represent a novel fold, the arrangement of His 143, Glu 163, and Cys 184 is similar to the active sites of other viral and cellular cysteine and serine proteases. The catalytic triad of $NS2^{pro}$ demonstrated a similar spatial distribution with the active sites from known cysteine or serine proteases (papain, poliovirus 3C protease, Sindbis virus capsid and subtilisin), suggesting that the composite active site formed by NS2 dimerization is competent for catalysis.[27] To the best of our knowledge, $NS2^{pro}$ represents the first example of a cysteine or serine protease with a composite active site that requires dimerization. However, those features are reminiscent of retroviral aspartic proteases, which consist of dimers with a single active site at the dimerization interface.[45] Other proteases such as caspases require dimerization for activity, but they do not contain composite active sites.[3]

Catalysis of a peptide bond by cysteine proteases proceeds through a well-characterized, charge relay network mechanism. The NS2 structure suggests that Glu 163 polarizes His 143 to deprotonate the cysteine side chain. The nucleophilic Cys 184 attacks the carbonyl of the scissile bond forming a negatively charged, tetrahedral transition state. Cleavage of the activated intermediate releases the N terminus of NS3 and creates an acyl-enzyme intermediate in which Cys 184 is covalently bonded to Leu 217. Hydrolysis of the acyl-enzyme intermediate forms the carboxylic acid of Leu 217 and restores the proton to Cys 184. Interestingly, Pro 164, which is adjacent to Glu 163 of the catalytic triad, has a *cis*-peptide conformation with the pyrrolidine ring lying on the same side as the carbonyl group of Glu 163 (Fig. 6). This proline residue is entirely conserved in HCV and the related GB virus sequences, implying that this position has an important structural and functional role. The proximity of Pro 164 to Glu 163 may bend the peptide backbone to establish the correct geometry of the glutamate side chain for catalysis. In addition, the *cis*-proline confirmation may contribute to dimer stabilization since the linker that connects the two subdomains follows Pro 164.

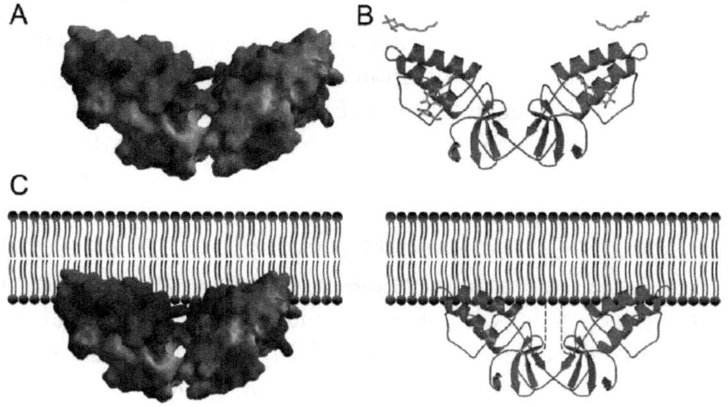

Figure 7. Model for membrane association of NS2pro. (A) Solvent-accessible surface of NS2pro dimer colored according to electrostatic potential. (B) Ribbon diagram of the NS2pro dimer with three n-octyl-β-glucoside and one n-decyl-β-maltoside molecules shown as a stick model. (C) Ribbon diagram and solvent-accessible surface relative to the ER membrane.

After cleavage, the C-terminal residue of NS2, Leu 217, remains bound to the active site. The carboxylic acid of Leu 217 contacts the side chains of the catalytic triad (His 143 and Cys 184) and the backbone nitrogen of Cys 184 (Fig. 6). This nitrogen represents the oxyanion hole that stabilizes the negatively charged transition state. There are few contacts between the side chain of Leu 217 and the rest of the protein, indicating that the cleavage site has minimal sequence requirements. This observation is consistent with published results showing that substitutions of amino acid residues near the scissile bond, including the C-terminal residue of NS2, have minor effects on cleavage efficiency.[34] We proposed a model in which binding of the C terminus to the active site precludes further proteolysis and thus catalyzes a single cleavage event per molecule. Although the data presented would indicate that the enzyme is catalytically inactive once the NS2/3 junction is cleaved, we cannot rule out the possibility that NS2 mediates a second proteolysis of other viral or cellular proteins. However, this would require displacing strand b5, which may disrupt the β sheet in the C-terminal subdomain, making a second cleavage event unlikely.

10. Membrane association of NS2

The solvent-accessible surface of $NS2^{pro}$ colored by electrostatic potential is shown in Fig. 7A. Generally, the molecule has a high content of neutral and basic regions, with a few acidic patches being present in the C-terminal subdomain. The surface of the molecule formed by the two α helices is mainly hydrophobic, with some basic residues lying underneath. Several molecules of n-octyl-β-glucoside and n-decyl-β-maltoside, which were used in the purification and crystallization, were located in the crystal structure of NS2 (Fig. 7B). The detergent molecules bind to the protein near helices H1 and H2. The hydrophobic electrostatic potential and the detergent binding sites suggest that the protease domain of NS2 may interact peripherally with the membrane of the ER. A model of the membrane association of NS2 is presented in Fig. 7C. The N-terminal end of helix H2 of each monomer would be inserted into the membrane and interact with the fatty acid tails of the membrane lipids, while basic residues in the N-terminal subdomain would neutralize the polar head groups. Peripheral membrane association of helix H2 would place the N termini of the two protein monomers in close proximity to the membrane. This is consistent with the putative topology model for the full-length NS2, in which the hydrophobic N-terminal third of NS2 contains several transmembrane segments.[36]

11. Evidence in favor of a NS2 dimer

Molecular surface analysis and biochemical data support the $NS2^{pro}$ dimer model. $NS2^{pro}$ shows a high degree of amino-acid sequence conservation at the interface between the two monomers (see supplementary figure 3 in ref [27]). The cleavage rate of purified NS2-3 is concentration dependent, indicating that the active form of the protease is oligomeric.[31] Analytical ultracentrifugation of $NS2^{pro}$ yielded a single, monodisperse species with a molecular weight of 39 kDa that most likely corresponds to a dimer (estimated MW of the dimer is about 29kDa) with bound detergent (data not shown). Moreover, cross-linking of $NS2^{pro}$ in solution with disuccinimidyl suberate (DSS) led to the identification of a dimeric species (Fig. 8).

A series of experiments in mammalian cells was designed to test whether $NS2^{pro}$ can form dimers with a functional composite active site *in vivo*. HCV full-length polyproteins containing either a H143A or a C184A mutation in the NS2 active site are defective in NS2-3 processing.[13,14] However, if a composite active site can form, co-expression of the two. mutant polyproteins should result in partial NS2-3 cleavage (see figure 3 in ref [27]).

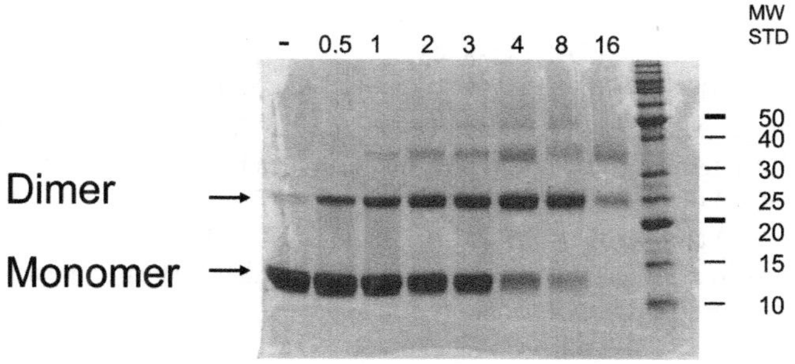

Figure 8. Chemical crosslinking of NS2[pro]. Purified NS2[pro] was incubated in the presence of DSS (Disuccinimidyl suberate), a non-cleavable, amine-reactive, homobifunctional cross-linker, separated on SDS-PAGE and stained with Coomassie blue. The molar excess of DSS to protein is shown across the top of the gel.

Indeed, when HCV polyproteins with NS2 containing either a H143A or C184A mutation were co-expressed, NS2 and NS3 cleavage products were detected, indicating the formation of a functional composite active site. These data strongly support the NS2[pro] crystal structure and prove that NS2 can form dimers with composite functional active sites.

References

1. Alberts, I. L., K. Nadassy, and S. J. Wodak. 1998. Analysis of zinc binding sites in protein crystal structures. *Protein Sci* 7:1700–16.
2. Blight, K., A. Kolykhalov, and C. Rice. 2000. Efficient initiation of HCV RNA replication in cell culture. Science 290:1972–4.
3. Boatright, K. M., and G. S. Salvesen. 2003. Mechanisms of caspase activation. *Curr Opin Cell Biol* 15:725–31.
4. Brass, V., E. Bieck, R. Montserret, B. Wolk, J. A. Hellings, H. E. Blum, F. Penin, and D. Moradpour. 2002. An amino-terminal amphipathic alpha-helix mediates membrane association of the hepatitis C virus nonstructural protein 5A. *J Biol Chem* 277:8130–9.
5. Brown, R. S. 2005. Hepatitis C and liver transplantation. Nature 436:973–8.
6. Choi, J., K. J. Lee, Y. Zheng, A. K. Yamaga, M. M. Lai, and J. H. Ou. 2004. Reactive oxygen species suppress hepatitis C virus RNA replication in human hepatoma cells. *Hepatology* 39:81–9.
7. Dimitrova, M., I. Imbert, M. P. Kieny, and C. Schuster. 2003. Protein–protein interactions between hepatitis C virus nonstructural proteins. *J Virol* 77:5401–14.
8. Elazar, M., K. H. Cheong, P. Liu, H. B. Greenberg, C. M. Rice, and J. S. Glenn. 2003. Amphipathic helix-dependent localization of NS5A mediates hepatitis C virus RNA replication. *J Virol* 77:6055–61.
9. Engh, R., and R. Huber. 1991. Accurate bond and angle parameters for X-ray protein structure refinement. *Acta Cryst.* A 47:337–45.

10. Evans, M. J., C. M. Rice, and S. P. Goff. 2004. Phosphorylation of hepatitis C virus nonstructural protein 5A modulates its protein interactions and viral RNA replication. *Proc Natl Acad Sci U S A* 101:13038–43.
11. Fried, M. W., M. L. Shiffman, K. R. Reddy, C. Smith, G. Marinos, F. L. Goncales, Jr., D. Haussinger, M. Diago, G. Carosi, D. Dhumeaux, A. Craxi, A. Lin, J. Hoffman, and J. Yu. 2002. Peginterferon alfa-2a plus ribavirin for chronic hepatitis C virus infection. *N Engl J Med* 347:975–82.
12. Gong, G., G. Waris, R. Tanveer, and A. Siddiqui. 2001. Human hepatitis C virus NS5A protein alters intracellular calcium levels, induces oxidative stress, and activates STAT-3 and NF-kappa B. *Proc Natl Acad Sci U S A* 98:9599–604.
13. Grakoui, A., D. W. McCourt, C. Wychowski, S. M. Feinstone, and C. M. Rice. 1993. A second hepatitis C virus-encoded proteinase. *Proc Natl Acad Sci U S A* 90:10583–7.
14. Hijikata, M., H. Mizushima, T. Akagi, S. Mori, N. Kakiuchi, N. Kato, T. Tanaka, K. Kimura, and K. Shimotohno. 1993. Two distinct proteinase activities required for the processing of a putative nonstructural precursor protein of hepatitis C virus. *J Virol* 67:4665–75.
15. Hijikata, M., H. Mizushima, Y. Tanji, Y. Komoda, Y. Hirowatari, T. Akagi, N. Kato, K. Kimura, and K. Shimotohno. 1993. Proteolytic processing and membrane association of putative nonstructural proteins of hepatitis C virus. *Proc Natl Acad Sci U S A* 90:10773–7.
16. Holm, L., and C. Sander. 1996. Mapping the protein universe. *Science* 273:595–603.
17. Huang, L., J. Hwang, S. D. Sharma, M. R. Hargittai, Y. Chen, J. J. Arnold, K. D. Raney, and C. E. Cameron. 2005. Hepatitis C virus nonstructural protein 5A (NS5A) is an RNA-binding protein. *J Biol Chem* 280:36417–28.
18. Jones, C. T., C. L. Murray, D. K. Eastman, J. Tassello, and C. M. Rice. 2007. Hepatitis C virus p7 and NS2 proteins are essential for production of infectious virus. *J Virol* 81:8374–83.
19. Jones, S., and J. M. Thornton. 1996. Principles of protein-protein interactions. *Proc Natl Acad Sci U S A* 93:13–20.
20. Kim, J. L., K. A. Morgenstern, C. Lin, T. Fox, M. D. Dwyer, J. A. Landro, S. P. Chambers, W. Markland, C. A. Lepre, E. T. O'Malley, S. L. Harbeson, C. M. Rice, M. A. Murcko, P. R. Caron, and J. A. Thomson. 1996. Crystal structure of the hepatitis C virus NS3 protease domain complexed with a synthetic NS4A cofactor peptide. *Cell* 87:343–55.
21. Kolykhalov, A. A., K. Mihalik, S. M. Feinstone, and C. M. Rice. 2000. Hepatitis C virus-encoded enzymatic activities and conserved RNA elements in the 3' nontranslated region are essential for virus replication in vivo. *J Virol* 74:2046–51.
22. Kubo, Y., K. Takeuchi, S. Boonmar, T. Katayama, Q. L. Choo, G. Kuo, A. J. Weiner, D. W. Bradley, M. Houghton, I. Saito, and et al. 1989. A cDNA fragment of hepatitis C virus isolated from an implicated donor of post-transfusion non-A, non-B hepatitis in Japan. *Nucleic Acids Res* 17:10367–72.
23. Leichert, L. I., and U. Jakob. 2004. Protein thiol modifications visualized in vivo. *PLoS Biol* 2:e333.
24. Li, P. P., A. Nakanishi, S. W. Clark, and H. Kasamatsu. 2002. Formation of transitory intrachain and interchain disulfide bonds accompanies the folding and oligomerization of simian virus 40 Vp1 in the cytoplasm. *Proc Natl Acad Sci U S A* 99:1353–8.
25. Lindenbach, B. D., Thiel, H. -J., Rice, C. M. 2007. Flavirviridae: The viruses and their replication. In D. M. Knipe (ed.), *Fields Virology*, 5th ed. Lippincott Williams & Wilkins, Philadelphia.
26. Lohmann, V., F. Korner, A. Dobierzewska, and R. Bartenschlager. 2001. Mutations in hepatitis C virus RNAs conferring cell culture adaptation. *J Virol* 75:1437–49.

27. Lorenz, I. C., J. Marcotrigiano, T. G. Dentzer, and C. M. Rice. 2006. Structure of the catalytic domain of the hepatitis C virus NS2-3 protease. *Nature* 442:831–5.
28. Love, R. A., H. E. Parge, J. A. Wickersham, Z. Hostomsky, N. Habuka, E. W. Moomaw, T. Adachi, and Z. Hostomska. 1996. The crystal structure of hepatitis C virus NS3 proteinase reveals a trypsin-like fold and a structural zinc binding site. *Cell* 87:331–42.
29. Macdonald, A., and M. Harris. 2004. Hepatitis C virus NS5A: tales of a promiscuous protein. *J Gen Virol* 85:2485–502.
30. Ostergaard, H., C. Tachibana, and J. R. Winther. 2004. Monitoring disulfide bond formation in the eukaryotic cytosol. *J Cell Biol* 166:337–45.
31. Pallaoro, M., A. Lahm, G. Biasiol, M. Brunetti, C. Nardella, L. Orsatti, F. Bonelli, S. Orru, F. Narjes, and C. Steinkuhler. 2001. Characterization of the hepatitis C virus NS2/3 processing reaction by using a purified precursor protein. *J Virol* 75:9939–46.
32. Penin, F., V. Brass, N. Appel, S. Ramboarina, R. Montserret, D. Ficheux, H. E. Blum, R. Bartenschlager, and D. Moradpour. 2004. Structure and function of the membrane anchor domain of hepatitis C virus nonstructural protein 5A. *J Biol Chem* 279:40835–43.
33. Qadri, I., M. Iwahashi, J. M. Capasso, M. W. Hopken, S. Flores, J. Schaack, and F. R. Simon. 2004. Induced oxidative stress and activated expression of manganese superoxide dismutase during hepatitis C virus replication: role of JNK, p38 MAPK and AP-1. *Biochem J* 378:919–28.
34. Reed, K. E., A. Grakoui, and C. M. Rice. 1995. Hepatitis C virus-encoded NS2-3 protease: cleavage-site mutagenesis and requirements for bimolecular cleavage. *J Virol* 69:4127–36.
35. Rho, J., S. Choi, Y. R. Seong, W. K. Cho, S. H. Kim, and D. S. Im. 2001. Prmt5, which forms distinct homo-oligomers, is a member of the protein-arginine methyltransferase family. *J Biol Chem* 276:11393–401.
36. Santolini, E., L. Pacini, C. Fipaldini, G. Migliaccio, and N. Monica. 1995. The NS2 protein of hepatitis C virus is a transmembrane polypeptide. *J Virol* 69:7461–71.
37. Shimakami, T., M. Hijikata, H. Luo, Y. Y. Ma, S. Kaneko, K. Shimotohno, and S. Murakami. 2004. Effect of interaction between hepatitis C virus NS5A and NS5B on hepatitis C virus RNA replication with the hepatitis C virus replicon. *J Virol* 78:2738–48.
38. Shirota, Y., H. Luo, W. Qin, S. Kaneko, T. Yamashita, K. Kobayashi, and S. Murakami. 2002. Hepatitis C virus (HCV) NS5A binds RNA-dependent RNA polymerase (RdRP) NS5B and modulates RNA-dependent RNA polymerase activity. *J Biol Chem* 277:11149–55.
39. Tan, S. L., and M. G. Katze. 2001. How hepatitis C virus counteracts the interferon response: the jury is still out on NS5A. *Virology* 284:1–12.
40. Tellinghuisen, T. L., J. Marcotrigiano, A. E. Gorbalenya, and C. M. Rice. 2004. The NS5A protein of hepatitis C virus is a zinc metalloprotein. *J Biol Chem* 279:48576–87.
41. Tellinghuisen, T. L., and C. M. Rice. 2002. Interaction between hepatitis C virus proteins and host cell factors. *Curr Opin Microbiol* 5:419–27.
42. Thibeault, D., R. Maurice, L. Pilote, D. Lamarre, and A. Pause. 2001. In vitro characterization of a purified NS2/3 protease variant of hepatitis C virus. *J Biol Chem* 276:46678–84.
43. Tu, H., L. Gao, S. T. Shi, D. R. Taylor, T. Yang, A. K. Mircheff, Y. Wen, A. E. Gorbalenya, S. B. Hwang, and M. M. Lai. 1999. Hepatitis C virus RNA polymerase and NS5A complex with a SNARE-like protein. *Virology* 263:30–41.
44. Weiner, A. J., G. Kuo, D. W. Bradley, F. Bonino, G. Saracco, C. Lee, J. Rosenblatt, Q. L. Choo, and M. Houghton. 1990. Detection of hepatitis C viral sequences in non-A, non-B hepatitis. *Lancet* 335:1–3.

45. Wlodawer, A., and A. Gustchina. 2000. Structural and biochemical studies of retroviral proteases. *Biochim Biophys Acta* 1477:16–34.
46. Wu, Z., N. Yao, H. V. Le, and P. C. Weber. 1998. Mechanism of autoproteolysis at the NS2-NS3 junction of the hepatitis C virus polyprotein. *Trends Biochem Sci* 23:92–4.
47. Zhang, J., O. Yamada, T. Sakamoto, H. Yoshida, T. Iwai, Y. Matsushita, H. Shimamura, H. Araki, and K. Shimotohno. 2004. Down-regulation of viral replication by adenoviral-mediated expression of siRNA against cellular cofactors for hepatitis C virus. *Virology* 320:135–43.

PROTEIN STRUCTURE MODELING

NARAYANAN ESWAR*, ANDREJ SALI*
Departments of Biopharmaceutical Sciences and Pharmaceutical Chemistry, California Institute for Quantitative Biosciences, University of California at San Francisco, San Francisco, CA, USA

Abstract. Known protein sequences outnumber known protein structures by more than two orders of magnitude. Given this huge sequence-structure gap, most protein structures need to be predicted by computational methods rather than determined by experimental techniques. This chapter outlines various protein structure modeling approaches and associated resources.

Keywords: Comparative modeling, homology modeling, threading, integrative modeling, sequence-structure alignment

1. Introduction

Cellular functions are dependent on the three-dimensional (3D) structures of proteins and their complexes with small molecules and other macromolecules. Knowing the structures of the proteins is thus crucial for the understanding of cellular processes. The 3D structures of the proteins and their complexes are best determined by experimental methods that yield solutions at atomic resolution, such as x-ray crystallography and nuclear magnetic resonance (NMR) spectroscopy. However, despite recent advances in the application of such techniques in high-throughput mode, experimental structural characterization remains an expensive and time-consuming task (Chandonia and Brenner, 2006).

* To whom correspondence should be addressed. Andrej Sali or Eswar Narayanan, UCSF MC 2552, Byers Hall Suite 503B, 1700 4th Street, San Francisco, CA 94158-2330, USA; e-mails: sali@salilab.org, eswar@salilab.org

The publicly available Protein Data Bank (PDB) currently contains only ~50,000 structures (Berman et al., 2007). In contrast, rapid improvements in genome sequencing resulted in approximately five million protein sequences, including the complete genetic blueprints of humans and hundreds of other organisms (Bairoch et al., 2005; Benson et al., 2005). This wide sequence-structure gap can only be bridged by computational means.

Fortunately, domains in protein sequences are evolving gradually and can thus be clustered into a relatively small number of families with similar sequences and structures (Vitkup et al. 2001; Chandonia and Brenner 2005b). For instance, ~80% of all sequences in the UniProt database can be clustered into approximately 10,000 families (Bru et al., 2005; Letunic et al., 2006; Finn et al., 2008). Similarly, all the structures in the PDB can be classified into approximately 1,000 distinct folds (Andreeva et al., 2004; Pearl et al., 2005). Many computational methods for protein structure modeling seek to exploit these evolutionary relationships.

Computational approaches to protein structure prediction are greatly facilitated by the structural genomics initiative (Liu et al., 2007; Moult, 2008). Structural genomics aims to maximize the structural coverage of the sequence space by experimentally determining the representative structures for as many families as possible, thus allowing accurate modeling of the remaining members of these families (Sali, 1998). Currently, most targets for experimental structure determination are chosen from the largest protein families such that, in combination with computational methods, each new structure yields useful structural information for the largest possible fraction of sequences in the shortest possible time frame (Chandonia and Brenner 2005a).

2. Computational structure determination

There are three main types of computational protein structure modeling methods. First, *ab initio* methods aim to predict the structure of a target protein purely from its primary sequence using principles of physics that govern protein folding and/or using information derived from known structures but without relying on any evolutionary relationship to known folds (Simons et al., 1999; Das and Baker, 2008). Currently, these methods can only be applied to individual domains of less than approximately 150 residues (Baker and Sali, 2001).

Second, homology or comparative modeling methods rely on the fact that similar sequences adopt similar 3D structures. Comparative modeling consists of four main steps (Marti-Renom et al., 2000): (i) fold assignment that identifies similarity between the target sequence of interest and at least one known protein structure (the template); (ii) alignment of the target sequence and the template(s); (iii) building an atomic model of the target based on the alignment with the chosen template(s); and (iv) predicting errors in the model. The first two steps, fold assignment and sequence-alignment, are frequently achieved by sequence-structure threading methods that seek to assess a coarse model derived by threading the target sequence through each structure within a library of protein folds (templates). The threading methods are most useful when the similarity between a target sequence and any of the known structures is not statistically significant (Godzik, 2003). Threading methods achieve higher sensitivity than sequence comparison methods by using structural information derived from the templates. Comparative modeling is the most accurate approach that can be easily applied on a large-scale to address the sequence-structure gap. Though these three classes of methods seem to address distinct regimes of the structure prediction problem, the divisions between them are increasingly being blurred. State-of-the-art modeling methods tend to employ the best features of each of these methods to improve the accuracy of the resulting models.

Finally, a third group of methods, recently receiving a lot of attention, is the "integrative" or "hybrid" methods that combine information from a varied set of computational and experimental sources, including those listed above (Alber et al., 2008).

3. Geometrical accuracy of comparative protein structure models

We now focus on comparative protein structure modeling. The geometrical accuracy of comparative models can be estimated by building models for sequences with known structures and comparing them to their native structures. Specifically, a measure of accuracy is usually plotted as a function of the sequence identity of the target-template alignment that was used to calculate the target model (Fig. 1).

Based on such comparisons, sequence-structure relationships are coarsely classified into three different regimes in the sequence similarity spectrum: (i) the easily detected relationships characterized by >30% sequence identity, (ii) the "twilight zone" (Rost, 1999) corresponding to relationships with statistically significant sequence similarity in the 10–30% range, and (iii) the "midnight zone" (Rost, 1999) corresponding to statistically insignificant sequence similarity – the regime where threading methods show the greatest promise.

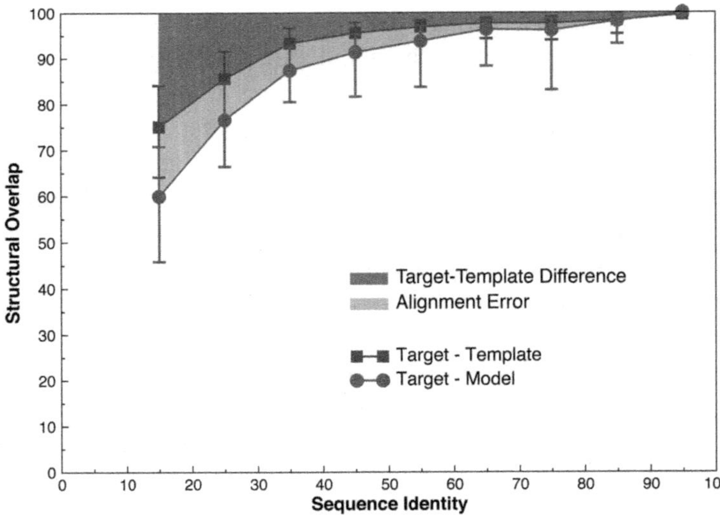

Figure 1. The median accuracy of comparative models plotted as a function of sequence identity. Structural overlap is defined as the fraction of equivalent C^{α} atoms. For the comparison of the model with the native structure (filled circles), two C^{α} atoms were considered equivalent if they belonged to the same residue and were within 3.5 Å of each other after least-squares superposition. For comparisons between the native structure and the template used for modeling (squares), two C^{α} atoms were considered equivalent if they were within 3.5 Å of each other after a structural alignment. The difference between the model and the actual target structure is a combination of the target-template differences (dark gray area) and the alignment errors (light gray area). The lower-panel indicates the most commonly seen model errors. The data was derived by analyzing approximately 1 million models produced by MODPIPE for sequences with known structures.

Models based on alignments with >30% sequence identity almost always have the correct fold. On average, such models also usually have >70–75% of the backbone atoms correctly modeled with a root-mean-squared-deviation (RMSD) of less than 3.5 Å (Fig. 1). However, as the sequence identity drops below 30%, even evolutionarily related proteins tend to show significant differences in their structures. These differences lead to errors in the alignment that, in turn, decrease the accuracy of the resulting model. Nevertheless, state-of-the-art alignment methods, including profile-sequence, profile–profile methods and structure-based environment dependent substitution matrices, have significantly improved the accuracy of such alignments (Shi et al., 2001; Wang and Dunbrack, 2004; Soding, 2005; Zhou and Zhou, 2005; Wu and Zhang, 2008). On average, it is not uncommon for models based on alignments with 20–30% sequence identity to have more than half the backbone atoms modeled accurately. For most alignments below 20% sequence identity, it still remains a challenge to calculate an accurate alignment. Getting a model that is close to the native structure, in this regime of sequence identity, involves exploring the conformational space without reliance on the alignment. There have been recent reports of success in addressing this problem (Bradley et al., 2005; Misura et al., 2006; Chen and Skolnick, 2008; Zhang, 2008). However, such approaches involve computationally expensive search strategies that prevent their application on a large-scale.

4. Prediction of model accuracy

The accuracy of the predicted model determines the information that can be extracted from it. Thus, estimating the accuracy of a model in the absence of the known structure is essential for its interpretation. As discussed above, a model calculated using a template structure that shares more than 30% sequence identity is indicative of an overall accurate structure (i.e., RMSD of the backbone atoms when compared to the native structure is within ~0.5–3.0 Å).

It is generally useful to assess errors in (i) the choice of template structures, (ii) the alignment, (iii) the modeling of loops, (iv) rigid-body shifts and distortions, and (v) the packing of side-chains. Thus, a number of assessment scores have been developed that specialize in evaluating specific aspects of protein structure models, such as: (i) determining whether or not a model has the correct fold (Tanaka and Scheraga, 1976; Sippl, 1993; Miyazawa and Jernigan, 1996; Domingues et al., 1999; Melo et al., 2002); (ii) discriminating between the native and near-native states (Lazaridis and Karplus, 1999; Gatchell et al., 2000; Vorobjev and Hermans, 2001; Tsai et al., 2003; Zhang et al., 2004; Shen and Sali, 2006); and (iii) selecting the most native-like model in a set of decoys that does not contain the native structure (Shortle et al., 1998; Eramian et al., 2006). Different measures to predict errors in a protein structure perform best at different levels of accuracy. For instance, physics-based force-fields may be helpful at identifying the best model when all models are very close to the native state (<1.5 Å RMSD over all backbone C^{α} atoms, corresponding to ~85% target-template sequence identity). In contrast, coarse-grained scores such as atomic distance-dependent statistical potentials have been shown to have the greatest ability to differentiate between models in the ~3 Å C^{α} RMSD range. Tests show that such scores are often able to identify a model within 0.5 Å C^{α} RMSD of the most accurate model produced (Eramian et al., 2006).

5. Evaluation of protein structure modeling methods

It is crucial for method developers and users alike to assess the accuracy of their methods. An attempt to address this problem has been made by the

CASP (Critical Assessment of Techniques for Proteins Structure Prediction) experiments (Kryshtafovych et al., 2005). These biannual competitions acquire experimentally determined protein structures before they are released to the public and allow participants to predict the structures, which are then evaluated by human experts. However, the major limitation of this competition is that it can assess methods only over a limited number of target protein sequences (Bujnicki et al., 2001; Marti-Renom et al., 2002) and only once every 2 years. To overcome these limitations, two additional evaluation experiments have been described, LiveBench (Bujnicki et al., 2001) and EVA (Eyrich et al., 2001; Koh et al., 2003), which continuously evaluate participating modeling web-servers over a cumulative period of time. For example, the aims of EVA are (i) to evaluate continuously and automatically blind predictions by prediction servers, based on identical and sufficiently large data sets; (ii) to provide weekly updates of the method assessments on the web; and (iii) to enable developers, non-expert users, and reviewers to determine the performance of the tested prediction servers.

6. Genome-scale protein structure modeling and databases

6.1. LARGE-SCALE PROTEIN STRUCTURE MODELING

There are several automated modeling methods, available through the internet, as evidenced by the increasing number of web-servers that participate in the CASP competitions. However, bridging the widening sequence-structure gap requires the development of completely automated, stable, reliable and, most importantly, scalable modeling methods than can be applied to millions of sequences. Currently, there are at least three such large-scale efforts that have been applied to entire genomes, including SWISS-MODEL (Schwede et al., 2003), MODPIPE (Eswar et al., 2003), and FAMS (Takeda-Shitaka et al., 2005). Results of such large-scale calculations indicate that it is currently possible to model at least one domain for over half of all the sequences in most genomes.

6.2. DATABASES OF PROTEIN STRUCTURE MODELS

Depositions to the PDB are restricted to atomic coordinates that are substantially determined by experimental measurements on specimens containing biological macromolecules (Berman et al., 2007). However, as mentioned above, several millions of comparative protein models have been generated for the protein sequences contained in the UniProtKB database using the experimentally determined structures in PDB. These models are disseminated to the community through individual databases such as MODBASE (Pieper et al. 2006), SWISS-MODEL REPOSITORY (Kopp and Schwede 2006), and FAMSBASE (Yamaguchi et al., 2003). Databases of annotated comparative models increase the efficiency for expert users, allow cross-referencing with other (non-structure-centric) resources, and make comparative models accessible to non-experts.

The Protein Model Portal (http://www.proteinmodelportal.org) has recently been developed as part of the PSI Structural Genomics Knowledge Base to provide an integrated access to the various databases containing structural information and thereby implementing the first step of the community workshop recommendation (Kouranov et al. 2006) on archiving structural models of biological macromolecules. Currently, models calculated by the six structural genomics centers, MODBASE, and SWISS-MODEL Repository are accessible through a single search interface.

7. Integrative or hybrid modeling techniques

Biological function cannot be provided by a single protein molecule in isolation. It is the result of stable or transient interactions among individual proteins and other molecules in the cell. Most of these interactions remain uncharacterized by traditional structural biology techniques such as X-ray crystallography and NMR spectroscopy. This gap is being bridged by several emerging experimental approaches that vary in terms of the information they provide (Robinson et al., 2007). For example, the stoichiometry and composition of protein components in an assembly can be determined by methods such as quantitative immunoblotting and mass spectrometry. The shape of the assembly can be revealed by electron microscopy and small angle X-ray scattering. The positions of the components can be elucidated by cryo-electron microscopy and labeling techniques. Whether or not components interact with each other can be measured by mass spectrometry, yeast two-hybrid and affinity purification. Relative orientations of components and information about interacting residues can be inferred from cryo-electron microscopy, hydrogen/deuterium exchange, hydroxyl radical footprinting, and chemical-crosslinking (Alber et al., 2008) (Fig. 2).

When approaches dominated by a single source of information fail, simultaneous consideration of all available information about the composition and structure of a given protein or assembly, irrespective of its source, can sometimes be sufficient to calculate a useful structural model (Robinson et al., 2007). Even when the model resulting from such integrative or hybrid methods is of relatively low resolution and accuracy, it can still be helpful for studying the function and evolution of the modelled protein or assembly; it also provides the necessary starting point for a higher resolution study.

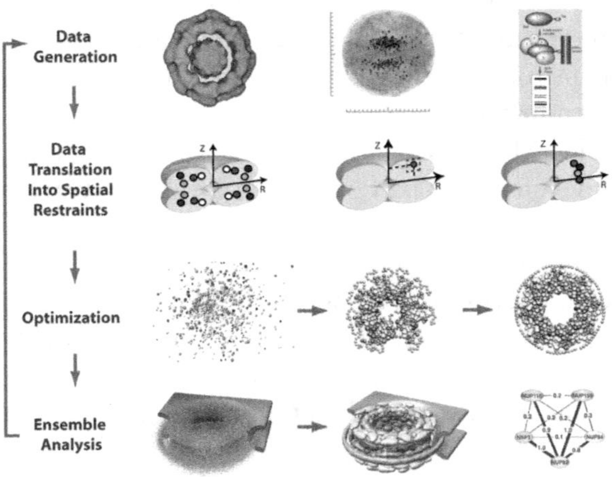

Figure 2. Integrative structure determination. The four steps of determining a structure of a protein or a macromolecular assembly by integration of varied data are illustrated with the example of the nuclear pore complex (Alber et al. 2007a, b; Robinson et al., 2007). First, structural data are generated by experiments, such as electron microscopy (left panel), immuno-electron microscopy (middle panel), and affinity purification of subcomplexes (right panel); many other types of information can also be added. Second, the data and theoretical considerations are expressed as spatial restraints ensuring the observed symmetry and shape of the assembly (electron microscopy, left panel), positions of constituent gold-labeled proteins (immuno-electron microscopy, middle panel), and proximity among the constituent proteins (affinity co-purification, right panel). Third, an ensemble of structural solutions that satisfy the data is obtained by minimizing the violations of the spatial restraints (from left to right). Fourth, the ensemble is clustered into sets of distinct solutions (left panel) as well as analyzed in different representations, such as protein positions (middle panel) and protein-protein contacts (right panel). The integrative approach to structure determination has several advantages: (i) It benefits from the synergy among the input data, minimizing the drawback of incomplete, inaccurate, and/or imprecise data sets (although each individual restraint may contain little structural information, the concurrent satisfaction of all restraints derived from independent experiments may drastically reduce the degeneracy of structural solutions); (ii) it can potentially produce all structures that are consistent with the data, not just one; (iii) the variation among the structures consistent with the data allows us to assess sufficiency of the data and the precision of the representative structure; (iv) it can make the process of structure determination more efficient by indicating what measurements would be the most informative. (This figure was reproduced from figure 5 inRobinson et al. (2007)).

An example of a simple hybrid approach is building a pseudo-atomic model of a large assembly by fitting atomic structures of subunits into its cryo-electron microscopy map (Gao et al., 2003; Chandramouli et al., 2008; Topf et al., 2008). X-ray diffraction data has been combined with protein structure modelling to provide solutions for molecular replacement (Qian et al., 2007). Unassigned or partially assigned NMR spectroscopy data and fragment-based modeling approaches have been combined to improve structure refinement in terms of its accuracy, efficiency, and success rate (Shen et al., 2008). A variety of different types of information, such as symmetry and protein proximity, have been used to characterize large symmetrical assemblies, including the nuclear pore complex (Alber et al., 2007b), EscJ from the type III secretion system (Andre et al., 2007), and the AAA+ ring complexes (Diemand and Lupas, 2006).

8. Future directions

8.1. PROTEIN STRUCTURE MODELING

Improvement in the accuracy of atomic comparative models will require methods that finely sample protein conformational space using a free energy or scoring function that has sufficient accuracy to distinguish the native structure from the non-native conformations. Despite many years of development of molecular simulation methods, attempts to refine models that are already relatively close to the native structure have met with relatively little success. This failure is likely to be due in part to inaccuracies in the scoring functions used in the simulations, particularly in the treatment of electrostatics and solvation effects. A combination of physics-based energy functions with the statistical information extracted from known protein structures may provide more accurate scoring functions. In addition to the scoring function, improvements in sampling strategies are also likely to be necessary.

8.2. INTEGRATIVE MODELING

Cryo-electron microscopy is emerging as a key technique for studying 3D structures of multi-component macromolecular complexes with masses larger than 250 kDa, such as membrane proteins, cytoskeletal complexes, ribosomes, quasi-spherical viruses, molecular chaperones, flagella, ion channels, and oligomeric enzymes. Electron cryo-tomography even enables the observation of macromolecules inside a living cell in its native state (Baumeister, 2004). Various modeling approaches are being developed that utilize cryo-electron microscopy density maps as a restraint in deriving a pseudo atomic model of the molecular components within a larger complex. Because of the significant likelihood of conformational differences between isolated domains and biological assemblies, additional research resulting in reliable hybrid modeling methods, which are able to correctly include structural information from various experimental sources of different resolution and reliability, is essential. Structural information from hybrid models, generating a synoptic image of the heterogeneous information available for a given macromolecular system, is expected to increase sharply in the coming years.

ACKNOWLEDGEMENTS

We thank Drs. Ben Webb, Mallur S Madhusudhan, Marc A Marti-Renom, Min-yi Shen, Ursula Pieper and David Eramian for helpful discussions about comparative modeling. This article is partially based on papers by Eswar and Sali (2007) and Schwede et al. (2008). We also acknowledge funds from Sandler Family Supporting Foundation, U.S. National Institutes of Health (Grants R01-GM54762, R01-GM083960, U54-RR022220, U54-GM074945, P01-GM71790, U54-GM074929), U.S. National Science Foundation (Grant IIS-0705196), as well as Hewlett-Packard, Sun Microsystems, IBM, NetApp Inc. and Intel Corporation for hardware gifts.

References

Alber, F., Dokudovskaya, S., Veenhoff, L.M., Zhang, W., Kipper, J., Devos, D., Suprapto, A., Karni-Schmidt, O., Williams, R., Chait, B.T., et al. 2007a. Determining the architectures of macromolecular assemblies. *Nature* **450:** 683–694.

Alber, F., Dokudovskaya, S., Veenhoff, L.M., Zhang, W., Kipper, J., Devos, D., Suprapto, A., Karni-Schmidt, O., Williams, R., Chait, B.T., et al. 2007b. The molecular architecture of the nuclear pore complex. *Nature* **450:** 695–701.

Alber, F., Forster, F., Korkin, D., Topf, M., and Sali, A. 2008. Integrating Diverse Data for Structure Determination of Macromolecular Assemblies. *Ann Rev Biochem.*

Andre, I., Bradley, P., Wang, C., and Baker, D. 2007. Prediction of the structure of symmetrical protein assemblies. *Proceedings of the Natl Acad Sci USA* **104:** 17656–17661.

Andreeva, A., Howorth, D., Brenner, S.E., Hubbard, T.J., Chothia, C., and Murzin, A.G. 2004. SCOP database in 2004: refinements integrate structure and sequence family data. *Nucleic Acids Res* **32:** D226–229.

Bairoch, A., Apweiler, R., Wu, C.H., Barker, W.C., Boeckmann, B., Ferro, S., Gasteiger, E., Huang, H., Lopez, R., Magrane, M., et al. 2005. The Universal Protein Resource (UniProt). *Nucleic Acids Res* **33:** D154–159.

Baker, D., and Sali, A. 2001. Protein structure prediction and structural genomics. *Science* **294:** 93–96.

Baumeister, W. 2004. Mapping molecular landscapes inside cells. *Biol Chem* **385:** 865–872.

Benson, D.A., Karsch-Mizrachi, I., Lipman, D.J., Ostell, J., and Wheeler, D.L. 2005. GenBank. *Nucleic Acids Res* **33:** D34–38.

Berman, H., Henrick, K., Nakamura, H., and Markley, J.L. 2007. The worldwide Protein Data Bank (wwPDB): ensuring a single, uniform archive of PDB data. *Nucleic Acids Res* **35:** D301–303.

Bradley, P., Misura, K.M., and Baker, D. 2005. Toward high-resolution de novo structure prediction for small proteins. *Science (New York)* **309:** 1868–1871.

Bru, C., Courcelle, E., Carrere, S., Beausse, Y., Dalmar, S., and Kahn, D. 2005. The ProDom database of protein domain families: more emphasis on 3D. *Nucleic Acids Res* **33:** D212–215.

Bujnicki, J.M., Elofsson, A., Fischer, D., and Rychlewski, L. 2001. LiveBench-1: continuous benchmarking of protein structure prediction servers. *Protein Sci* **10:** 352–361.

Chandonia, J.M., and Brenner, S. 2005a. Update on the pfam5000 strategy for selection of structural genomics targets. *Conf Proc IEEE Eng Med Biol Soc* **1:** 751–755.

Chandonia, J.M., and Brenner, S.E. 2005b. Implications of structural genomics target selection strategies: Pfam5000, whole genome, and random approaches. *Proteins* **58:** 166–179.

Chandonia, J.M., and Brenner, S.E. 2006. The impact of structural genomics: expectations and outcomes. *Science (New York)* **311:** 347–351.

Chandramouli, P., Topf, M., Menetret, J.F., Eswar, N., Cannone, J.J., Gutell, R.R., Sali, A., and Akey, C.W. 2008. Structure of the Mammalian 80S Ribosome at 8.7 A Resolution. *Structure* **16:** 535–548.

Chen, H., and Skolnick, J. 2008. M-TASSER: an algorithm for protein quaternary structure prediction. *Biophys J* **94:** 918–928.

Das, R., and Baker, D. 2008. Macromolecular Modeling with Rosetta. *Ann Rev Biochem*.

Diemand, A.V., and Lupas, A.N. 2006. Modeling AAA+ ring complexes from monomeric structures. *J Structur Biol* **156:** 230–243.

Domingues, F.S., Koppensteiner, W.A., Jaritz, M., Prlic, A., Weichenberger, C., Wiederstein, M., Floeckner, H., Lackner, P., and Sippl, M.J. 1999. Sustained performance of knowledge-based potentials in fold recognition. *Proteins* **Suppl 3:** 112–120.

Eramian, D., Shen, M.Y., Devos, D., Melo, F., Sali, A., and Marti-Renom, M.A. 2006. A composite score for predicting errors in protein structure models. *Protein Sci* **15:** 1653–1666.

Eswar, N., and Sali, A. 2007. Comparative modeling of drug target proteins. In *Comprehensive Medicinal Chemistry II*. (ed. J.S. Mason), pp. 215–236. Elsevier, Oxford.

Eswar, N., John, B., Mirkovic, N., Fiser, A., Ilyin, V.A., Pieper, U., Stuart, A.C., Marti-Renom, M.A., Madhusudhan, M.S., Yerkovich, B., et al. 2003. Tools for comparative protein structure modeling and analysis. *Nucleic Acids Res* **31:** 3375–3380.

Eyrich, V.A., Marti-Renom, M.A., Przybylski, D., Madhusudhan, M.S., Fiser, A., Pazos, F., Valencia, A., Sali, A., and Rost, B. 2001. EVA: continuous automatic evaluation of protein structure prediction servers. *Bioinformatics* **17:** 1242–1243.

Finn, R.D., Tate, J., Mistry, J., Coggill, P.C., Sammut, S.J., Hotz, H.R., Ceric, G., Forslund, K., Eddy, S.R., Sonnhammer, E.L., et al. 2008. The Pfam protein families database. *Nucleic Acids Res* **36:** D281–288.

Gao, H., Sengupta, J., Valle, M., Korostelev, A., Eswar, N., Stagg, S.M., Van Roey, P., Agrawal, R.K., Harvey, S.C., Sali, A., et al. 2003. Study of the structural dynamics of the E coli 70S ribosome using real-space refinement. *Cell* **113:** 789–801.

Gatchell, D.W., Dennis, S., and Vajda, S. 2000. Discrimination of near-native protein structures from misfolded models by empirical free energy functions. *Proteins* **41:** 518–534.

Godzik, A. 2003. Fold recognition methods. *Methods Biochem Anal* **44:** 525–546.

Koh, I.Y., Eyrich, V.A., Marti-Renom, M.A., Przybylski, D., Madhusudhan, M.S., Eswar, N., Grana, O., Pazos, F., Valencia, A., Sali, A., et al. 2003. EVA: Evaluation of protein structure prediction servers. *Nucleic Acids Res* **31:** 3311–3315.

Kopp, J., and Schwede, T. 2006. The SWISS-MODEL Repository: new features and functionalities. *Nucleic Acids Res* **34:** D315–318.

Kouranov, A., Xie, L., de la Cruz, J., Chen, L., Westbrook, J., Bourne, P.E., and Berman, H.M. 2006. The RCSB PDB information portal for structural genomics. *Nucleic Acids Res* **34:** D302–305.

Kryshtafovych, A., Venclovas, C., Fidelis, K., and Moult, J. 2005. Progress over the first decade of CASP experiments. *Proteins* **61 Suppl 7:** 225–236.

Lazaridis, T., and Karplus, M. 1999. Discrimination of the native from misfolded protein models with an energy function including implicit solvation. *J Mol Biol* **288:** 477–487.

Letunic, I., Copley, R.R., Pils, B., Pinkert, S., Schultz, J., and Bork, P. 2006. SMART 5: domains in the context of genomes and networks. *Nucleic Acids Res* **34:** D257–260.

Liu, J., Montelione, G.T., and Rost, B. 2007. Novel leverage of structural genomics. *Nature Biotechnol* **25:** 849–851.

Marti-Renom, M.A., Stuart, A.C., Fiser, A., Sanchez, R., Melo, F., and Sali, A. 2000. Comparative protein structure modeling of genes and genomes. *Annu Rev Biophys Biomol Struct* **29:** 291–325.

Marti-Renom, M.A., Madhusudhan, M.S., Fiser, A., Rost, B., and Sali, A. 2002. Reliability of assessment of protein structure prediction methods. *Structure (Camb)* **10:** 435–440.

Melo, F., Sanchez, R., and Sali, A. 2002. Statistical potentials for fold assessment. *Protein Sci* **11:** 430–448.

Misura, K.M., Chivian, D., Rohl, C.A., Kim, D.E., and Baker, D. 2006. Physically realistic homology models built with ROSETTA can be more accurate than their templates. *Proc Natl Acad Sci USA* **103:** 5361–5366.

Miyazawa, S., and Jernigan, R.L. 1996. Residue-residue potentials with a favorable contact pair term and an unfavorable high packing density term, for simulation and threading. *J Mol Biol***256:** 623–644.

Moult, J. 2008. Comparative modeling in structural genomics. *Structure* **16:** 14–16.

Pearl, F., Todd, A., Sillitoe, I., Dibley, M., Redfern, O., Lewis, T., Bennett, C., Marsden, R., Grant, A., Lee, D., et al. 2005. The CATH Domain Structure Database and related resources Gene3D and DHS provide comprehensive domain family information for genome analysis. *Nucleic Acids Res* **33:** D247–251.

Pieper, U., Eswar, N., Davis, F.P., Braberg, H., Madhusudhan, M.S., Rossi, A., Marti-Renom, M., Karchin, R., Webb, B.M., Eramian, D., et al. 2006. MODBASE: a database of annotated comparative protein structure models and associated resources. *Nucleic Acids Res* **34:** D291–295.

Qian, B., Raman, S., Das, R., Bradley, P., McCoy, A.J., Read, R.J., and Baker, D. 2007. High-resolution structure prediction and the crystallographic phase problem. *Nature* **450:** 259–264.

Robinson, C.V., Sali, A., and Baumeister, W. 2007. The molecular sociology of the cell. *Nature* **450:** 973–982.

Rost, B. 1999. Twilight zone of protein sequence alignments. *Protein Eng* **12:** 85–94.

Sali, A. 1998. 100,000 protein structures for the biologist. *Nat Struct Biol* **5:** 1029–1032.

Schwede, T., Kopp, J., Guex, N., and Peitsch, M.C. 2003. SWISS-MODEL: An automated protein homology-modeling server. *Nucleic Acids Res* **31:** 3381–3385.

Schwede, T., Sali, A., Eswar, N., and Peitsch, M.C. 2008. Protein Structure Modeling. In *Computational Structural Biology*. (eds. T. Schwede, and M.C. Peitsch). World Scientific Publishing, Singapore.

Shen, M.Y., and Sali, A. 2006. Statistical potential for assessment and prediction of protein structures. *Protein Sci* **15:** 2507–2524.

Shen, Y., Lange, O., Delaglio, F., Rossi, P., Aramini, J.M., Liu, G., Eletsky, A., Wu, Y., Singarapu, K.K., Lemak, A., et al. 2008. Consistent blind protein structure generation from NMR chemical shift data. *Proc Natl Acad Sci USA* **105:** 4685–4690.

Shi, J., Blundell, T.L., and Mizuguchi, K. 2001. FUGUE: sequence-structure homology recognition using environment-specific substitution tables and structure-dependent gap penalties. *J Mol Biol* **310:** 243–257.

Shortle, D., Simons, K.T., and Baker, D. 1998. Clustering of low-energy conformations near the native structures of small proteins. *Proc Natl Acad Sci USA* **95:** 11158–11162.

Simons, K.T., Bonneau, R., Ruczinski, I., and Baker, D. 1999. Ab initio protein structure prediction of CASP III targets using ROSETTA. *Proteins* **Suppl 3:** 171–176.

Sippl, M.J. 1993. Boltzmann's principle, knowledge-based mean fields and protein folding. An approach to the computational determination of protein structures. *J Comp-Aid Mol Design* **7:** 473–501.

Soding, J. 2005. Protein homology detection by HMM-HMM comparison. *Bioinformatics (Oxford, England)* **21:** 951–960.

Takeda-Shitaka, M., Terashi, G., Takaya, D., Kanou, K., Iwadate, M., and Umeyama, H. 2005. Protein structure prediction in CASP6 using CHIMERA and FAMS. *Proteins* **61 Suppl 7:** 122–127.

Tanaka, S., and Scheraga, H.A. 1976. Medium- and long-range interaction parameters between amino acids for predicting three-dimensional structures of proteins. *Macromolecules* **9:** 945–950.

Topf, M., Lasker, K., Webb, B., Wolfson, H., Chiu, W., and Sali, A. 2008. Protein structure fitting and refinement guided by cryo-EM density. *Structure* **16:** 295–307.

Tsai, J., Bonneau, R., Morozov, A.V., Kuhlman, B., Rohl, C.A., and Baker, D. 2003. An improved protein decoy set for testing energy functions for protein structure prediction. *Proteins* **53:** 76–87.

Vitkup, D., Melamud, E., Moult, J., and Sander, C. 2001. Completeness in structural genomics. *Nat Struct Biol* **8:** 559–566.

Vorobjev, Y.N., and Hermans, J. 2001. Free energies of protein decoys provide insight into determinants of protein stability. *Protein Sci* **10:** 2498–2506.

Wang, G., and Dunbrack, R.L., Jr. 2004. Scoring profile-to-profile sequence alignments. *Protein Sci* **13:** 1612–1626.

Wu, S., and Zhang, Y. 2008. MUSTER: Improving protein sequence profile-profile alignments by using multiple sources of structure information. *Proteins.*

Yamaguchi, A., Iwadate, M., Suzuki, E., Yura, K., Kawakita, S., Umeyama, H., and Go, M. 2003. Enlarged FAMSBASE: protein 3D structure models of genome sequences for 41 species. *Nucleic Acids Res* **31:** 463–468.

Zhang, Y. 2008. I-TASSER server for protein 3D structure prediction. *BMC Bioinforma.* **9:** 40.

Zhang, C., Liu, S., Zhou, H., and Zhou, Y. 2004. An accurate, residue-level, pair potential of mean force for folding and binding based on the distance-scaled, ideal-gas reference state. *Protein Sci* **13:** 400–411.

Zhou, H., and Zhou, Y. 2005. SPARKS 2 and SP3 servers in CASP6. *Proteins* **61 Suppl 7:** 152–156.

STRUCTURAL BIOLOGY AND MOLECULAR MODELING IN THE DESIGN OF NOVEL DPP-4 INHIBITORS

GIOVANNA SCAPIN*
Departments of Global Structural Biology, Merck & Co., Inc., PO BOX 2000, Rahway, NJ 07065, USA

Abstract. Inhibition of dipeptidyl peptidase IV (DPP-4) is a promising new approach for the treatment of type 2 diabetes. DPP-4 is the enzyme responsible for inactivating the incretin hormones glucagon-like peptide 1 (GLP-1) and glucose dependent insulinotropic polypeptide (GIP), two hormones that play important roles in glucose homeostasis. The potent, orally bioavailable and highly selective small molecule DPP-4 inhibitor sitagliptin has been approved by the FDA as novel drug for the treatment of type 2 diabetes. The comparison between the binding mode of sitagliptin (a β-amino acid) and that of a second class of inhibitors (α-amino acid-based) initially led to the successful identification and design of structurally diverse and highly potent DPP-4 inhibitors. Further analysis of the crystal structure of sitagliptin bound to DPP-4 suggested that the central β-amino butanoyl moiety could be replaced by a rigid group. This was confirmed by molecular modeling, and the resulting cyclohexylamine analogs were synthesized and found to be potent DPP-4 inhibitors. However, the triazolopyrazine was predicted to be distorted in order to fit in the binding pocket, and the crystal structure showed that multiple conformations exist for this moiety. Additional molecular modeling studies were then used to improve potency of the cyclohexylamine series. In addition, a 3-D QSAR method was used to gain insight for reducing off-target DPP-8/9 activities. Novel compounds were thus synthesized and found to be potent DPP-4 inhibitors. Two compounds in particular were designed to be highly selective against off-target "DPP-4 Activity- and/or Structure Homologues" (DASH) enzymes while maintaining potency against DPP-4.

---

* To whom correspondence should be addressed. Giovanna Scapin, Departments of Global Structural Biology, Merck & Co., Inc., PO BOX 2000, Rahway, NJ 07065, USA. Current address: Schering-Plough Research Institute, 2015 Galloping Hill Road K15-1-1800, Kenilworth, NJ 07033, USA; e-mail: giovanna.scapin@spcorp.com

J.L. Sussman and P. Spadon (eds.), *From Molecules to Medicines*,

Keywords: Diabetes, dipeptidyl peptidase, selectivity, structural biology, molecular modeling

1. Introduction

1.1. DIABETES

Type 2 diabetes is a component of a complex metabolic syndrome, characterized by abnormal insulin secretion caused by impaired β-cell function and insulin resistance in target tissues.[1,2] It has emerged as one of the world's most debilitating diseases, and its prevalence is reaching epidemic proportions, with over 300 million cases expected worldwide by 2025.[3] The pathogenesis of type 2 diabetes is complex and it is influenced by both genetic and environmental factors, including obesity and physical inactivity.[4] The medical and socioeconomic burden of the disease is caused by the associated complications[5–9] and is imposing enormous strains on health-care systems. The landmark symptom of type 2 diabetes is elevated blood glucose levels, caused by several processes, including improper insulin production in the pancreas, improper glucose uptake and insulin utilization in the muscles and improper glucose production in the liver. Several treatments are currently available for the management of hyperglycemia, including metformin, thiazolidinediones, sulfonylurea derivatives and exogenous insulin, but they all have serious problems associated with them, for example, weight gain, the potential of occasionally severe hypoglycemia, and the lack of long term efficacy.[10,11] There clearly is the need for the identification of new therapeutic targets that could address these issues.

1.2. DPP-4

Dipeptidyl peptidase 4 (E.C. 3.4.14.5, DPP-4 or CD26) is a multifunctional glycoprotein with a serine dipeptidase activity, present both in circulation and on the cell surface. The membrane bound DPP-4 is tethered by an N-terminal peptide of about 35 residues that upon cleavage yields the soluble form.[12] Since the early 1990s, DPP-4 has been known to play a major role in regulating the inactivation of glucagon-like peptide 1 (GLP-1)[13,14] and glucose-dependent insulinotropic polypeptide (GIP)[15] in vitro and in vivo. GLP-1 and GIP are responsible for several actions, including glucose-induced insulin biosynthesis and secretion, inhibition of glucagon secretion, regulation of gene expression, slowing gastric emptying and trophic effects on β-cells. These effects are related to the normalization of the blood glucose level, and control of appetite and body weight (for lead DPP-4 references and its physiological roles see Conarello et al.[16] DPP-4 inhibition increases circulating GLP-1 and GIP levels in humans, which lead to decreased blood glucose levels, hemoglobin A_{1c} levels, and glucagon levels.(for reviews of available data, see refs [17–19]). DPP-4 inhibitors were also shown to offer a number of advantages over existing diabetes therapies including a lowered risk of hypoglycemia, the potential for weight loss, and the potential for the regeneration and differentiation of pancreatic β-cells.[20,21] Nevertheless, there were several concerns regarding potential safety issues for DPP-4 inhibition, including effects on immune cells[22] and regulation of several peptides in vivo.[23] DPP-4 is a member of a larger family of "DPP-4 activity- and/or structure-homologues" (DASH) enzymes, characterized by the common cleavage of a peptide bond found after a proline 24. This family includes the quiescent cell proline dipeptidase (QPP or DPP7),[25,26] DPP8,[27] DPP9,[28] and fibroblast activation protein (FAP).[29] Except for DPP-IV, the functions of these enzymes are unknown, but given the common mechanism, they are likely to be involved in some of the biological processes that are regulated by proline-specific amino-terminal processes.[30] Thus, although reports that DPP-4 deficient mice developed normally and were healthy[16,31] indicated that the enzyme was indeed a good target, the fact that inhibiting one or more of these other DASH enzymes could have unknown consequences suggested that the potential for off-target toxicity remained an issue that needed to be addressed in the development of suitable compounds.

1.3. OFF-TARGET ACTIVITY

At least one of the first compounds reported in human clinical trials (compound **1a**, L-*threo*-isoleucyl thiazolide 32), exhibited severe toxicities in chronic toxicity studies in dogs[33] which led to discontinuation of development of the compound in 2001.[34] Interestingly, of the two isomers of **1b**, one (the *allo* isomer) was 10-fold more toxic to rats and dogs than the other (the *threo* isomer), despite having similar pharmacodynamics and pharmacokinetics in both species. This suggested that the toxicity observed was likely not related to DPP-4 inhibition, but to some off-target activity.

To address the issue of the importance of selectivity in the design of DPP-4 inhibitors, the *allo* and *threo* compounds were tested against, among other proteases, the DASH enzymes. The *allo*-compound (the more toxic one) was found to be about five- to tenfold more potent against DPP8 and DPP9 than the *threo* compound, in agreement with the observed differences in potency required to produce toxicity.[33] Based on these facts, it was hypothesized that DPP8/9 inhibition was responsible for the observed toxicity,[33] additional data were also consistent with these observations, and that prompted for the search of a highly selective DPP-4 inhibitor.

2. The structure of DPP-4

The structure of human DPP-4 was initially reported in 2003 as a complex with the pseudo substrate Val-Pyr[35]; within the next two months, four more structures were reported, a human, apo-structure,[36] the complexes with diprotin A[37] and a covalent inhibitor,[38] and the dipeptide-complexed porcine kidney structure.[39] In 2004, the structure of DPP-4 in complex with a deca-peptide derived from the physiological substrate neuropeptide Y was also reported.[40] Since then, several other structures of the soluble form of DPP-4 in complex with different substrates and inhibitors have been reported, thus providing further evidences for the understanding of compound potency and selectivity.

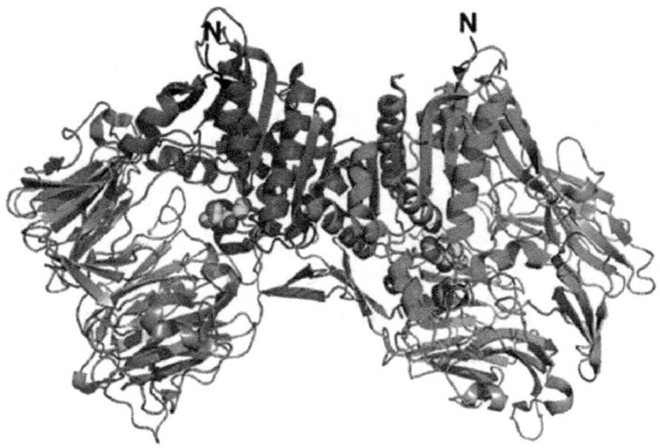

Figure 1. DPP-4 dimer (from PDB entry 1N1M): DPP-4 is a homodimer (red and green subunit); each subunit contains an α/β hydrolase domain (top, dark color) and a β-propeller domain (bottom, light color). The inhibitor Val-pyr, shown as ball-and-stick model, is at the interface of the two domains.

2.1. OVERALL STRUCTURE OF DPP-4

DPP-4 was found to be a dimer in all crystals reported to date, in agreement with biological data that the active enzyme is a dimer[41] (Fig. 1). The soluble form of DPP-4 lacks the first 38 amino-acids, which are the transmembrane peptide that anchors DPP-4 to the cell-membrane. In the crystal structures, the N-terminal end of each subunit is located on the same side of the complex, suggesting that the full-length enzyme could exist as a dimer even when bound to the membrane.

Each subunit contains two domains: an α/β hydrolase domain and an eight-bladed β-propeller domain. The binding site, originally identified by the presence of the competitive inhibitor valine-pyrrolidide (Val-Pyr) molecule, is located in a smaller pocket within the large cavity formed by the two domains. The serine protease active triad (Ser630, Asp708 and His740) is located within this pocket, as are the other residues now know to be important for binding and catalytic activity, including Arg125, Glu205, Glu206, Tyr662, Tyr666 and Asn710[35,40,42] (Fig. 2). A tyrosine residue found near the catalytic pocket, Tyr547, has been proposed to be an essential part of the catalytic mechanism of DPP-4,[43] and also one of the players that mediate compounds selectivity for DPP-4 over DPP8/9 44. This is probably due to differences in conformational flexibility of this conserved residue, which would allow for specific inhibitors to be active against DPP-4 but not DPP8/9.

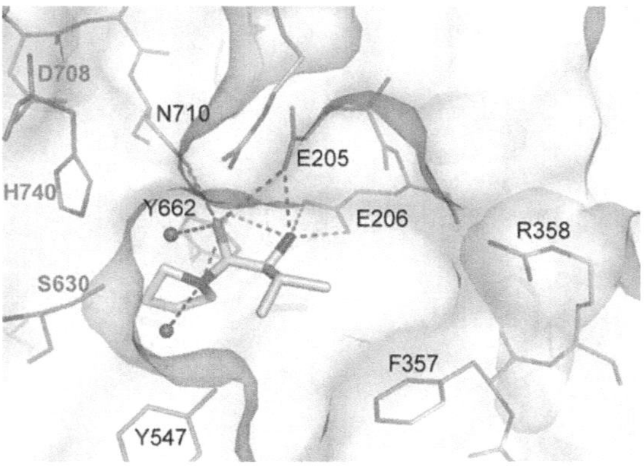

Figure 2. DPP-4 binding site (from PDB file 1N1M). The Val-Pyr molecule is shown with yellow carbons. The catalytic triad is colored in cyan.

2.2. THE STRUCTURE OF α- AND β-AMINO ACID DPP-4 INHIBITORS

Among the several classes of DPP-4 inhibitors that have been developed in the recent years (for review, see refs 19,20), only two will be discussed here in structural terms, since they are the precursors for the structure-guided inhibitor design described in this paper.

2.3. α-AMINO ACID INHIBITORS

Initial work done in the Merck laboratories converted the modestly potent and poorly selective compound **1a** into potent and selective β-methylphenylalanine compounds with the general structure **2.**[45,46] Further optimization of the β-substituted phenylalanine compounds derived led to the discovery of the potent and selective compound **3**.[47] This compound incorporated many of the characteristics that have been discovered to be essential for compounds that were potent and selective and retained excellent pharmacokinetic profiles. The structure of compound **3** bound to DPP-4 indicates that the compound binds as an amino acid substrate (Fig. 3). The fluoropyrrolidine overlays with the proline ring of the Val-pyr molecule, and the carbonyl and amino groups make the same interactions with Glu205, Glu206, Tyr662 and Asn710. Compound **3** extends much further into the DPP-4 binding pocket, and interacts with the side chain of Arg358. The triazolopyridine is stacked on top, and is within van deer Waals distance of the side chain of Phe357; these hydrophobic interactions probably contribute to the high potency of Compound **3**. A hydrogen bond is formed with the side chain of Tyr547, a residue that has been shown to be important for DPP-4 activity[43] and perhaps selectivity.[44]

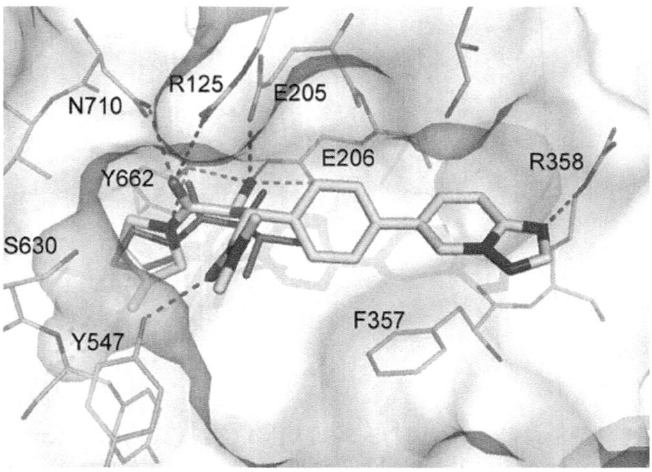

Figure 3. Compound **3** bound to DPP-4; the Val-Pyr molecule (magenta) is shown for comparison

2.3.1. *β-Amino-acid inhibitors*

Screening of the Merck sample collection led to two different classes of inhibitors, piperazine **4** and β-amino amide **5**. Extensive SAR on the right and left-hand side of the molecules eventually produced the 14 nM DPP-4 inhibitor **6.**[48] In order to stabilize the metabolically labile piperazine ring, chemistry focused on substituted bicyclic piperazine replacements, which provided improved metabolism and bioavailability. Adjustments of the fluorine substituents provided increased potency, and eventually sitagliptin (compound **7**) was synthesized.[49] Since there was no structure of the β-amino acid inhibitors bound to DPP-4, their binding mode with respect to the α-amino acid series was not known, but SAR studies suggested that it might be reversed. While only small changes were tolerated on the amide substituent of the α-amino acids, large amide modifications were tolerated on the same region in the β-amino acid series. On the other hand, large side chains were allowed for the α-amino acids, but only small variations of the phenyl ring were tolerated for the β-amino acids.[34]

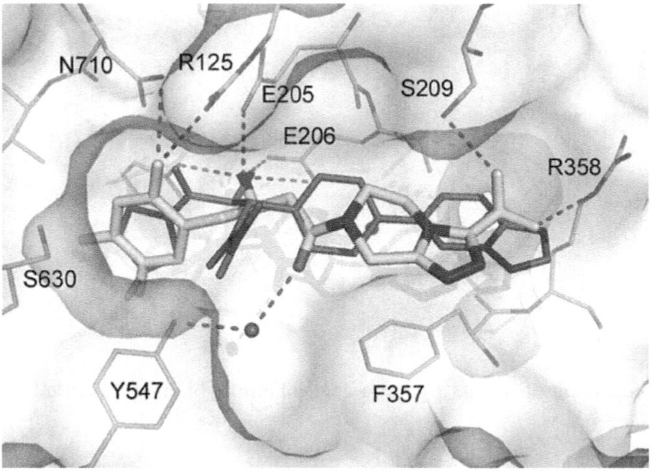

Figure 4. Binding of sitagliptin to DPP-4. Compound 3 (magenta) is shown for comparison.

When the structure of sitagliptin bound to DPP-4 was solved,[49] it confirmed that the binding was indeed reversed (Fig. 4). Crystal structures of all the other compounds in the series showed similar binding modes (G. Scapin, unpublished results). The reverse binding allows the β-amino acid inhibitors to retain the same interactions that the α-amino acid compounds have, specifically the interactions between the amino group and the side chains of Glu205, Glu206, and Tyr662; one of the fluorine atoms in the trifluorophenyl ring replaces the carboxylate, and interacts with Asn710 and Arg125. As in the α-amino acid series, the compound extends much further into the DPP-4 binding pocket, and interacts with the side chains of Phe357 (hydrophobic interaction) and Ser209 and Arg358 (polar interactions). A water molecule bridges the compound to the side chain of Tyr547.

3. Crystal structures, molecular modeling and rational design

3.1. INITIAL MODELING WORK: α-SERIES VERSUS β-SERIES

Once the binding mode of both series was established, efforts were started to use the structural information available for the design of structurally novel and diverse inhibitors that would nevertheless retain potency and selectivity. As indicated in Fig. 5, comparison of the binding mode of α- and β-amino acid compounds suggested that the two classes had some shared functionality, in particular the presence of conserved hydrogen bond acceptors and carbon atoms. Based on this observations, two new series of compounds were synthesized, exemplified by compounds **6** and **7** (G. Liang, manuscript in preparation).

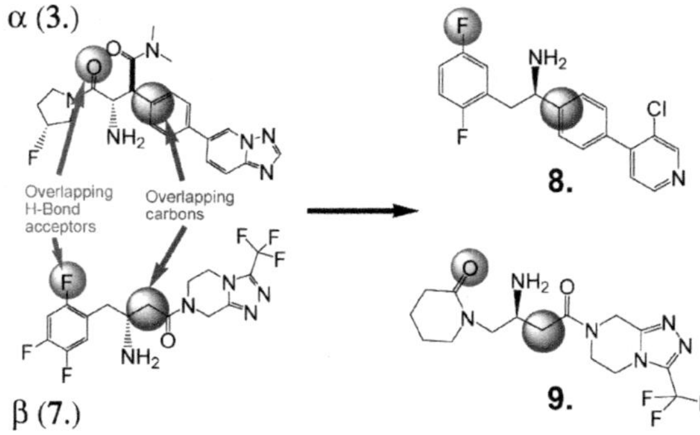

Figure 5. Comparison of α- and β-amino acid inhibitors binding mode led to the design of compounds **8**. and **9.**

Compound **8** in particular was very exciting, because it represented the first attempt to include structural information in the design of novel compounds. The IC_{50} of compound **8** was ~100 nM, and this not only provided the chemists with a good starting point for the development of a potential new lead, but also confirmed that the rational approach was working. Crystal structures of compounds in the two series showed that they bound as predicted, and that the features incorporated from the α- and β-series were positioned as expected

3.2. FROM SITAGLIPTIN TO CYCLOHEXYLAMINE INHIBITORS

The structure of sitagliptin (compound **7**) contains a β-amino butanoyl group as linker between the trifluorophenyl and the triazolopiperazine. We were interested in replacing this flexible moiety with a rigid analog, and when the structure of **7** bound to DPP-4 became available it was suggested, and later confirmed by molecular modeling that a cyclohexylamine could represent a good replacement for the β-amino butanoyl group. Based on this observation, compound **10** was synthesized and evaluated for potency and efficacy, and found to have similar potency to sitagliptin, excellent pharmacokinetic properties and excellent profile in the Oral Glucose tolerance test.[50]

The structure of **10** bound to DPP-4 (Fig. 6, overlaid onto sitagliptin) also confirmed that the cyclohexylamine was a good replacement for the β-amino butanoyl group. Interestingly, during the modeling work, it was suggested that the triazolopiperazine moiety, in the newly designed compound, might be distorted in order to fit into the binding site. Indeed, in the crystal structure, multiple conformations were observed for the compound, although it was possible to clearly assign only two (Fig. 8). In one of the conformers, the triazolopiperazine is stacked over the side chain of Phe357 (as sitagliptin does), and the trifluoromethyl is loosely interacting with the side chains of Ser209 and Arg358. In the second conformer, the triazolo-piperazine is perpendicular to the side chain of Phe357, and one of the nitrogens is bridged, through a water molecule, to the side chain of His126 and the main chain oxygen of Glu206. This observation suggested that the triazolopiperazine may not be the best substituent in this series, and prompted the search for better replacements.[51]

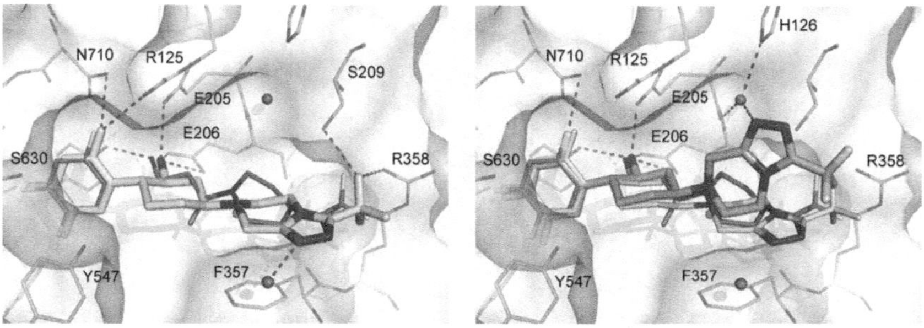

Figure 6. Left: compound **10** bound to DPP-4; sitagliptin (magenta) is also displayed for comparison. Right: two conformers were observed in the DPP-4:compound **10** complex.

3.3. FROM CYCLOHEXYLAMINES TO PYRROLOPYRIMIDINE INHIBITORS

Based on the structure of **10**, molecular modeling suggested that a 5–6 fused ring (isoindane) would probably provide a better fit to the DPP-4 binding pocket than the 6–5 fused ring (triazolopiperazine) used in **10**, thus increasing the compound potency 51. This compound (**11**) was synthesized and evaluated for potency against DPP-4 and selectivity against other members of the DASH family. Compound **11** gave (as predicted) a threefold boost in potency compared with **10**, and the crystal structure of a related compound (**12**, Fig. 7, see below) confirmed the binding mode predicted by modeling: there is no disorder in binding, and the isoindane p-stacks with the side chain of Phe357 better than compound **10** does. Unfortunately, **11** also displayed substantial activity against other DASH enzymes, especially DPP-8, DPP-9 and FAP 51. This was a concern, since inhibition of DPP-8 and/or DPP-9 is associated with severe toxicity in preclinical species.[33]

To address the selectivity issue, and in absence of structural information for DPP-8/9, molecular modeling undertook a 3D-QSAR approach that analyzed the structures of almost 3,000 compounds with IC_{50}s less than 500 nM for DPP-4 and available data for DPP-8. The resulting contour plots from the analysis suggested that heteroatoms or polar substituents on the heterocycle would help to increase the DPP-4 IC_{50}/DPP-8 IC_{50} ratio, thus providing more selective DPP-4 inhibitors.[51] Analogs with a pyrimidine moiety (**12**, **13**, and **14,** Table1) were synthesized and tested. In all cases, the potency remained equal to or greater than the one measured for **11**; in addition, the DPP-4 IC_{50}/DPP-8 IC_{50} ratio was greater that 7,500-fold, while for compound **11** it was 258-fold.

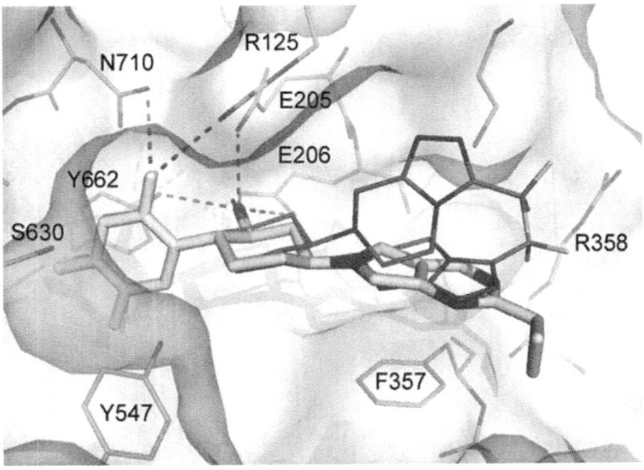

Figure 7. Compound **12** bound to DPP-4; Compound **12** (both conformations) is also displayed for comparison.

TABLE 1. Potency (IC50) and selectivity (DPP-8/DPP-4 ratio) of selected pyrrole-pyrimidine compounds.

Compound	DPP-4 (nM)	DPP-8 (μM)	Ratio
11.	6.6	1.7	258
12.	5.3	45.6	8,600
13.	0.3	4.2	14,000
14.	0.67	5.0	7,400

The fact that very similar results had been obtained for all three compounds suggests that it was the replacement with the pyridine moiety that contributed significantly to the selectivity, and not the introduction of the cyclo-propyl or trifluoromethyl substituents. Nevertheless, the substituents do appear to play a role in the compounds potency (there is an almost 18-fold increase in potency for 13 compared to 12. Crystal structures of the compounds 12 and 13 bound to DPP-4 show that they basically overlay (Fig. 8, left), but the trifluoromethyl provides a much better complementarity to the binding site than the cyclepropyl ring (Fig. 8 right), thus explaining the difference in potency.

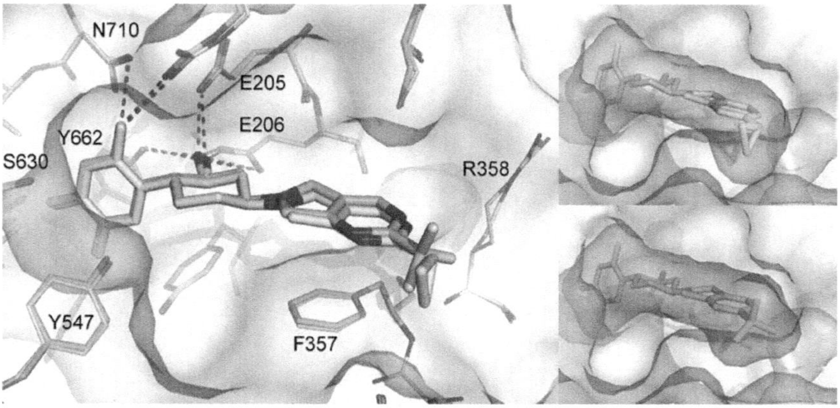

Figure 8. Compounds **12** (yellow) and **13** (green) bound to DPP-4:Left: overlay of the two compounds in the binding site; Right: surface representation of **12** (*top*) and **13** (*bottom*).

3.4. A FEW WORDS ABOUT SELECTIVITY

As of today, there are no crystal structures available for DPP-8 and DPP-9, and thus direct comparison is not possible. Nevertheless, some explanations for the selectivity observed for different classes of inhibitors have been suggested, based on structural and modeling work. The structure of the rat kidney DPP-4 in complex with different cyanopyrrolidine inhibitors 44 led to the suggestion that the conformation flexibility of Tyr547 may be a DPP-8/9 selectivity determinant. Although this residue is conserved in DPP8 and DPP9, modeling suggested it did not have the same conformational mobility, due to the replacement of the nearby Ser553 with a bulkier valine.[44] Indeed, compounds that cause shifting of this tyrosine side chain are shown to be selective.[44] Rummey and Metz have recently published homology models of DPP-8 and DPP-9, based on the crystal structure of DPP-4[52]. Analysis of the binding mode of sitagliptin to DPP-4 and comparison with the predicted binding sites in DPP-8 and DPP-9 led to the suggestion that the high selectivity displayed by sitagliptin may be due not to steric clashes with any of the subsites, but rather to the absence of key interactions in DPP-8 and DPP-9. The most important missing contacts correspond to the interactions with Ser209, Phe357, and Arg358, which belongs to two loops characterized by low homology and different conformation. Since all of the compounds described in this paper retain at least the interaction with Phe357, the same line of reasoning can be used to explain their selectivity, although other effects (such as desolvation costs and different distribution of charges within the binding site) may also play a role in defining selectivity.

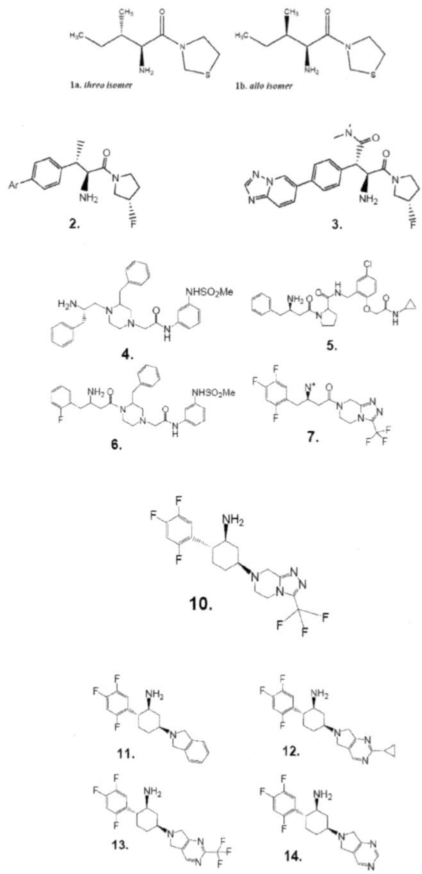

Figure 9. Chemical structures of the DPP-4 inhibitors described in the text.

ACKNOWLEDGMENTS

Many people have been involved in the DPP-4 program at Merck. In particular, I would like to thanks Xiaoping Zhang, Joseph Wu and Barbara Leiting for providing the protein, Sangita Patel for crystallizing it, and Suresh Singh and Ying-Duo Gao for the modeling work. Particular thanks to Ann Weber and Nancy Thornberry for continuous support and undisputed leadership and to all the talented chemists I worked with. Many thanks to the staff of the Industrial Macromolecular Crystallography Association Collaborative Access Team (IMCA-CAT) at the Advanced Photon Source for support during data collection. Use of the APS was supported by the U.S. Department of Energy, Basic Energy Sciences. Office of Science, under Contract No. W-31-109-Eng-38.

References

1. Taylor, S.I., "Deconstructing type 2 diabetes". *Cell*, **97**, 9–12 (1999).
2. Miranda, P.J., DeFronzo, Ralph, Califf, Robert M. and Guyton, John R., "Metabolic syndrome: Definition, pathophysiology, and mechanisms". *Am. Heart J.*, **149**, 33–45 (2004).
3. Diabetes Atlas 2003 [Available from: http://www.eatlas.idf.org/prevalence].
4. Hussain, A., Claussen, B., Ramachandran, A., and Williams, R., "Prevention of type 2 diabetes: a review". *DiabetesRres. Clinic. Pract.*, **76**, 317–326 (2007).
5. Zimmet, P., Alberti, K.G. and Shaw, J., "Global and societal implications of the diabetes epidemic.". *Nature*, **414**, 782–787 (2001).
6. Mensah, G.A., Mokdad, A.H., Ford, E., Narayan, K.M., Giles, W.H., Vinicor, F., and Deedwania, P.C., "Obesity, metabolic syndrome, and type 2 diabetes: emerging epidemics and their cardiovascular implications." *Cardiol. Clin.*, **22**, 485–504 (2004).
7. Paoletti, R., Bolego, C., Poli, A., and Cignarella, A., "Metabolic syndrome, inflammation and atherosclerosis.". *Vasc Health Risk Manag.*, **2**, 193–194 (2006).
8. Bray, G.A., and Bellanger, T., "Epidemiology, trends, and morbidities of obesity and the metabolic syndrome." *Endocrine*, **29**, 109–117 (2006).
9. Ceska, R., "Clinical implications of the metabolic syndrome." *Diab. Vasc. Dis. Res.*, **4**, S2–4 (2007).
10. Panunti, B., Jawa, Ali A., and Fonseca, Vivian A., "Mechanisms and therapeutic targets in type 2 diabetes mellitus". *Drug Discovery Today: Disease Mechanisms*, **1**, 151–157 (2004).
11. Stumvoll, M., Goldstein, B.J., and van Haeften, T.W., "Type 2 diabetes: principles of pathogenesis and therapy". *Lancet*, **365**, 1333–1346 (2005).
12. Lambeir, A.-M., Durinx, C., Scharpe, S., and De Meester, I., "Dipeptidyl-Peptidase IV from Bench to Bedside: An Update on Structural Properties, Functions, and Clinical Aspects of the Enzyme DPP IV". *Crit. Rev. Clin. Lab. Sci.*, **40**, 209–294 (2003).
13. Mentlein, R., Gallwitz, B., and Schmidt, W.E., "Dipeptidyl-peptidase IV hydrolyses gastric inhibitory polypeptide, glucagon-like peptide-1(7-36)amide, peptide histidine methionine and is responsible for their degradation in human serum.". *Eur. J. Biochem.*, **214**, 829–835 (1993).
14. Deacon, C.F., Johnsen, A.H., and Holst, J.J., "Degradation of glucagon-like peptide-1 by human plasma in vitro yields an N-terminally truncated peptide that is a major endogenous metabolite in vivo." *J. Clin. Endocrinol. Metab.* **80**, 952–957 (1995).
15. Kieffer, T.J., McIntosh, C.H., and Pederson, R.A., "Degradation of glucose-dependent insulinotropic polypeptide and truncated glucagon-like peptide 1 in vitro and in vivo by dipeptidyl peptidase IV." *Endocrinology*, **136**, 3585–3586 (1995).
16. Conarello, S.L., Li, Z., Ronan, J., Roy, R.S., Zhu, L., Jiang, G., Liu, F., Woods, J., Zycband, E., Moller, D.E., Thornberry, N.A., and Zhang, B.B., "Mice lacking dipeptidyl peptidase IV are protected against obesity and insulin resistance". *Proc Natl Acad Sci U S A*, **100**, 6825–6830 (2003).
17. Deacon, C.F., "Therapeutic Strategies Based on Glucagon-Like Peptide 1". *Diabetes*, **53**, 2181–2189 (2004).
18. Zerilli, T., and Pyon, E. Y., "Sitagliptin phosphate: A DPP-4 inhibitor for the treatment of type 2 diabetes mellitus". *Clin Ther*, **29**, 2614–2634 (2007).
19. Deacon, C.F., Carr, R.D., and Holst, J.J., "DPP-4 inhibitor therapy: new directions in the treatment of type 2 diabetes". *Front Biosci*, **13**, 1780–1794 (2008).
20. Ahren, B., "DPP-4 inhibitors". *Best Pract Res Clin Endocrinol Metab*, **21**, 517–533 (2007).

21. Pratley, R.E., and Salsali, A., "Inhibition of DPP-4: a new therapeutic approach for the treatment of type 2 diabetes". *Curr Med Res Opin.*, **23**, 919–931 (2007).
22. Mentlein, R., "Dipeptidyl-peptidase IV (CD26)–role in the inactivation of regulatory peptides". *Regul Pept.*, **85**, 9–24 (1999).
23. Reinhold, D., Kahne, T., Steinbrecher, A., Wrenger, S., Neubert, K., Ansorge, S., and Brocke, S., "The role of dipeptidyl peptidase IV (DP IV) enzymatic activity in T cell activation and autoimmunity". *Biol. Chem.*, **383**, 1133–1138 (2002).
24. Rosenblum, J.S. and Kozarich, J.W., "Prolyl peptidases: a serine protease subfamily with high potential for drug discovery". *Curr. Opin. Chem. Biol.*, **7**, 496–504 (2003).
25. Chiravuri, M., Schmitz, T., Yardley, K., Underwood, R., Dayal, Y., and Huber, B.T., "A novel apoptotic pathway in quiescent lymphocytes identified by inhibition of a post-proline cleaving aminodipeptidase: a candidate target protease, quiescent cell proline dipeptidase". *J.Immun. (Baltimore, Md : 1950)*, **163**, 3092–3099 (1999).
26. Underwood, R., Chiravuri, M., Lee, H., Schmitz, T., Kabcenell, A.K., Yardley, K., and Huber, B.T., "Sequence, purification, and cloning of an intracellular serine protease, quiescent cell proline dipeptidase". *J. Biol. Chem.*, **274**, 34053–34058 (1999).
27. Abbott, C.A., Yu, D.M., Woollatt, E., Sutherland, G.R., McCaughan, G.W., and Gorrell, M.D., "Cloning, expression and chromosomal localization of a novel human dipeptidyl peptidase (DPP) IV homolog, DPP8". *Euro J. Biochem. / FEBS*, **267**, 6140–6150 (2000).
28. Olsen, C. and Wagtmann, N., "Identification and characterization of human DPP9, a novel homologue of dipeptidyl peptidase IV". *Gene*, **299**, 185–193 (2002).
29. Scanlan, M.J., Raj, B.K., Calvo, B., Garin-Chesa, P., Sanz-Moncasi, M.P., Healey, J.H., Old, L.J., and Rettig, W.J., "Molecular cloning of fibroblast activation protein alpha, a member of the serine protease family selectively expressed in stromal fibroblasts of epithelial cancers". *Proc. Natl. Acad. Sci.USA*, **91**, 5657–5661 (1994).
30. Chen, W.-T., Kelly, T., and Ghersi, G., "DPPIV, seprase, and related serine peptidases in multiple cellular functions". *Curr.Ttopics Develop.Biol.*, **54**, 207–232 (2003).
31. Marguet, D., Baggio, L., Kobayashi, T., Bernard, A.M., Pierres, M., Nielsen, P.F., Ribel, U., Watanabe, T., Drucker, D.J., and Wagtmann, N., "Enhanced insulin secretion and improved glucose tolerance in mice lacking CD26". *Proc. Natl. Acad. Sci.USA*, **97**, 6874–6879 (2000).
32. Sorbera, L.A., Revel, L., and Castañer, J., "P32/98, Antidiabetic, Dipeptidyl-peptidase IV inhibitor." *Drugs. Fut.*, **26**, 859–864 (2001).
33. Lankas, G.R., Leiting, B., Roy, R.S., Eiermann, G.J., Beconi, M.G., Biftu, T., Chan, C.-C., Edmondson, S., Feeney, W.P., He, H., Ippolito, D.E., Kim, D., Lyons, K.A., Ok, H.O., Patel, R.A., Petrov, A.N., Pryor, K.A., Qian, X., Reigle, L., Woods, A., Wu, J.K., Zaller, D., Zhang, X., Zhu, L., Weber, A.E., and Thornberry, N.A., "Dipeptidyl Peptidase IV Inhibition for the Treatment of Type 2 Diabetes: Potential Importance of Selectivity Over Dipeptidyl Peptidases 8 and 9". *Diabetes*, **54**, 2988–2994 (2005).
34. Thornberry, N.A. and Weber, A.E., "Discovery of JANUVIA (TM) (Sitagliptin), a selective dipeptidyl peptidase IV inhibitor for the treatment of type2 diabetes". *Curr. Topics Med. Chem.*, **7**, 557–568 (2007).
35. Rasmussen, H.B., Branner, S., Wiberg, F.C., and Wagtmann, N., "Crystal structure of human dipeptidyl peptidase IV/CD26 in complex with a substrate analog". *Nat Struct Biol*, **10**, 19–25 (2003).
36. Hiramatsu, H., Kyono, K., Higashiyama, Y., Fukushima, C., Shima, H., Sugiyama, S., Inaka, K., Yamamoto, A., and Shimizu, R., "The structure and function of human dipeptidyl peptidase IV, possessing a unique eight-bladed beta-propeller fold". *Biochem. Biophys. Res. Comm.* **302**, 849–854 (2003).

37. Thoma, R., Loffler, B., Stihle, M., Huber, W., Ruf, A., and Hennig, M., "Structural basis of proline-specific exopeptidase activity as observed in human dipeptidyl peptidase-IV". *Structure*, **11**, 947–959 (2003).
38. Oefner, C., D'Arcy, A., Mac Sweeney, A., Pierau, S., Gardiner, R. and Dale, G E., "High-resolution structure of human apo dipeptidyl peptidase IV/CD26 and its complex with 1-[({2-[(5-iodopyridin-2-yl)amino]-ethyl}amino)-acetyl]-2-cyano -(S)-pyrrol idine." *Acta Crystallogr D Biol Crystallogr*, **D59**, 1206–1212 (2003).
39. Engel, M., Hoffmann, T., Wagner, L., Wermann, M., Heiser, U., Kiefersauer, R., Huber, R., Bode, W., Demuth, H.U., and Brandstetter, H., "The crystal structure of dipeptidyl peptidase IV(CD26) reveals its functional regulation and enzymatic mechanism". *Proc Natl Acad Sci U S A*, **100**, 5063–5068 (2003).
40. Aertgeerts, K., Ye, S., Tennant, M.G., Kraus, M. L., Rogers, J., Sang, B.-C., Skene, R.J., Webb, D.R., and Prasad, G.S., "Crystal structure of human dipeptidyl peptidase IV in complex with a decapeptide reveals details on substrate specificity and tetrahedral intermediate formation". *Prot. Sci* **13**, 412–421 (2004).
41. Gorrell, M.D., Gysbers, V., and McCaughan, G.W., "CD26: a multifunctional integral membrane and secreted protein of activated lymphocytes". *Scand. J. Immunol.*, **54**, 249–264 (2001).
42. Kopcho, L.M., Kim, Y.B., Wang, A., Liu, M.A., Kirby, M.S., and Marcinkeviciene, J., "Probing prime substrate binding sites of human dipeptidyl peptidase-IV using competitive substrate approach". *Arch Biochem Biophys*, **436**, 367–376 (2005).
43. Bjelke, J.R., Christensen, J., Branner, S., Wagtmann, N., Olsen, C., Kanstrup, A.B., and Rasmussen, H.B., "Tyrosine 547 Constitutes an Essential Part of the Catalytic Mechanism of Dipeptidyl Peptidase IV". *J Biol Chem*, **279**, 34691–34697 (2004).
44. Longenecker, K.L., Stewart, K.D., Madar, D.J., Jakob, C.G., Fry, E.H., Wilk, S., Lin, C.W., Ballaron, S.J., Stashko, M.A., Lubben, T.H., Yong, H., Pireh, D., Pei, Z., Basha, F., Wiedeman, P.E., von Geldern, T.W., Trevillyan, J.M., and Stoll, V.S., "Crystal structures of DPP-IV (CD26) from rat kidney exhibit flexible accommodation of peptidase-selective inhibitors". *Biochemistry*, **45**, 7474–7482 (2006).
45. Xu, J., Wei, L., Mathvink, R., He, J., Park, Y.-J., He, H., Leiting, B., Lyons, K.A., Marsilio, F., Patel, R.A., Wu, J.K., Thornberry, N. A., and Weber, A. E., "Discovery of potent and selective phenylalanine based dipeptidyl peptidase IV inhibitors". *Bioorg. Med. Chem. Lett.* **15**, 2533–2536 (2005).
46. Edmondson, S.D., Mastracchio, Duffy, J.L., Eiermann, G.J., He, H., Ita, I., Leiting, B., Leone, J.F., Lyons, K.A., Makarewicz, A.M., Patel, R.A., Petrov, A., Wu, J.K., Thornberry, N.A., and Weber, A.E., "Discovery of potent and selective orally bioavailable β-substituted phenylalanine derived dipeptidyl peptidase IV inhibitors". *Bioorg. Med. Chem. Lett*, **15**, 3048–3052 (2005).
47. Edmondson, S.D., Mastracchio, A., Mathvink, R.J., He, J.F., Harper, B., Park, Y.J., Beconi, M., Di Salvo, J., Eiermann, G.J., He, H.B., Leiting, B., Leone, J.F., Levorse, D.A., Lyons, K., Patel, R.A., Patel, S.B., Petrov, A., Scapin, G., Shang, J., Roy, R.S., Smith, A., Wu, J.K., Xu, S.Y., Zhu, B., Thornberry, N.A., and Weber, A.E., "(2S, 3S)-3-Amino-4-(3,3-difluoropyrrolidin-1-yl)-N,N-dimethyl4-oxo-2-(4- 1,2, 4 triazolo 1,5-a pyridin-6-ylphenyl)butanamide: A selective alpha-amino amide dipeptidyl peptidase IV inhibitor for the treatment of type 2 diabetes". *J. Med. Chem.*, **49**, 3614–3627 (2006).
48. Brockunier, L.L., He, J., Colwell, Jr., L.F., Habulihaz, B., He, H., Leiting, B., Lyons, K.A., Marsilio, F., Patel, R.A., Teffera, Y., Wu, J.K., Thornberry, N.A., Weber, A.E., and Parmee, E.R., "Substituted piperazines as novel dipeptidyl peptidase IV inhibitors". *Bioorg. Med. Chem. Lett*, **14**, 4763–4766 (2004).

49. Kim, D., Wang, L., Beconi, M., Eiermann, G.J., Fisher, M.H., He, H., Hickey, G.J., Kowalchick, J.E., Leiting, B., Lyons, K., Marsilio, F., McCann, M.E., Patel, R.A., Petrov, A., Scapin, G., Patel, S.B., Roy, R.S., Wu, J.K., Wyvratt, M.J., Zhang, B.B., Zhu, L., Thornberry, N.A., and Weber, A.E., "(2R)-4-oxo-4-[3-(trifluoromethyl)-5,6-dihydro[1,2,4]triazolo[4,3-a]pyrazin-7(8H)-yl]-1-(2,4,5-trifluorophenyl)butan-2-amine:a potent, orally active dipeptidyl peptidase IV inhibitor for the treatment of type 2 diabetes". *J. Med.Chem.*, **48**, 141–151 (2005).
50. Biftu, T., Scapin, G., Singh, S., Feng, D., Becker, J.W., Eiermann, G., He, H., Lyons, K., Patel, S., Petrov, A., Sinha-Roy, R., Zhang, B., Wu, J., Zhang, X., Doss, G.A., Thornberry, N.A., and Weber, A.E., "Rational design of a novel, potent, and orally bioavailable cyclohexylamine DPP-4 inhibitor by application of molecular modeling and X-ray crystallography of sitagliptin". *Bioorg. Med. Chem. Lett.,* **17**, 3384–3387 (2007).
51. Gao, Y.D., Feng, D., Sheridan, R.P., Scapin, G., Patel, S.B., Wu, J.K., Zhang, X.P., Sinha-Roy, R., Thornberry, N.A., Weber, A.E., and Biftu, T., "Modeling assisted rational design of novel, potent, and selective pyrrolopyrimidine DPP-4 inhibitors". *Bioorg. Med. Chem. Lett.,* **17**, 3877–3879 (2007).
52. Rummey, C., and Metz, G., "Homology models of dipeptidyl peptidases 8 and 9 with a focus on loop predictions near the active site". *PROTEINS: Struct., Function Bioinformat.* **66**, 160–171 (2007).

TOOLS TO MAKE 3D STRUCTURAL DATA MORE COMPREHENSIBLE: EMOVIE & PROTEOPEDIA

ERAN HODIS[1,2], JAIME PRILUSKY[2,3],
JOEL L. SUSSMAN[1,2*]
[1]*Department of Structural Biology, Weizmann Institute of Science, Rehovot, Israel*
[2]*The Israel Structural Proteomics Center, Weizmann Institute of Science, Rehovot, Israel*
[3]*Biological Services Unit, Weizmann Institute of Science, Rehovot, Israel*

Abstract. To the crystallographer, solving a three-dimensional (3D) protein or molecular structure often times feels like the ultimate success, and surely it is. However, of utmost importance is the communication of the insights revealed by the 3D structure, especially those insights that relate structure to function. In order for these insights to reach their potential for guiding future research, they must reach biologists. The problem is that 3D structures are inherently complex and thus communicating insights about 3D structures to non-structural biologists can be difficult. To aid the structural biologist in this endeavor, we have created two useful tools. The first, *eMovie*, is a plug-in for PyMOL that makes creating macromolecular animations much more simple. The second, *Proteopedia*, is a community-annotated 'wiki' web-resource that links descriptive text to 3D views of structures, resulting in intuitive communication of structural information.

Keywords: Communicating structural biology, 3D, animations, movies, dissemination, wiki, eMovie, Proteopedia, education, instruction

* To whom correspondence should be addressed. Joel L. Sussman, Department of Structural Biology, Weizmann Institute of Science, Rehovot 76100, Israel; e-mail: Joel.Sussman@weizmann.ac.il

1. Introduction

We present here two tools to aid in the communication of structural biology. One is a tool called eMovie[1] that facilitates the creation of molecular animations. The other is a large-scope, 'wiki' web resource called Proteopedia[2] that has a powerful capability for communicating three-dimensional (3D) structures by linking descriptive text to 3D views of the structure.

They are disparate tools, and thus we present them in separate sections, but it should be noted that movies created using eMovie can be uploaded to Proteopedia to aid in the description of a particular protein, molecule, or concept.

2. eMovie

2.1. ABSTRACT

The 3D structures of macromolecules are difficult to grasp and also to communicate. By their nature, movies or animations are particularly useful for highlighting key features by offering a 'guided tour' of structures and conformation changes. However, high-quality movies are rarely seen because they are currently difficult and time consuming to make. By adopting the traditional movie 'storyboard' concept, which gives guidance and direction to filming, eMovie makes the creation of lengthy molecular animations much easier. This tool is a plug-in for the open-source molecular graphics program PyMOL, and enables experts and novices alike to produce informative and high-quality molecular animations.

2.2. INTRODUCTION

Soon after the first 3D structures of biological macromolecules were determined, it became clear that visualization and comparison of these structures is crucial to understand their structure and function. The field of molecular graphics was pioneered by Cyrus Levinthal in the 1960s,[3] the '3D effect' being achieved by rotating the macromolecular structure constantly on the screen, like a 'real-time' movie. The largest drawback was that the computer system – a specialized and extremely large Digital Equipment Computer dedicated to this task, nicknamed the 'Kluge' – cost well over US$300,000. However, during the past 40 years, both hardware and software have progressed enormously (http://www.umass.edu/microbio/rasmol/history.htm) making it possible to accomplish much more on a standard desktop computer than was ever possible using the Kluge. A few notable examples of molecular visualization programs that have taken advantage of the improved technology include Kinemage,[4] the RasMol-based[5] 3DBrowser[6,7] and Protein Explorer (http://www.umass.edu/microbio/chime/explorer) and the Jmol-based (http://www.jmol.org) FirstGlance in Jmol (http://molvis.sdsc.edu/fgij).

Here, we present 'eMovie' – a tool for making the process of molecular movie creation more fluid and natural. It is more similar to traditional movie-editing programs and can be used to create extended, complex animations. eMovie introduces a storyboard to the world of molecular animation in addition to modular, insertable actions. Thus, the user can focus on the scientific story rather than the technicalities of animation.

eMovie has been created as a plug-in for the molecular display program PyMOL. PyMOL combines superior picture quality with a highly advanced scripting interface, including versatile animation commands that can create long, complex movies; however, accessing the animation commands through the command line is difficult and error-prone. A new user is faced with a steep learning curve given the complicated animation commands and the necessity for writing external scripts. eMovie resolves these issues by providing a simple graphical interface through which the user can interact with the powerful movie-making capabilities of PyMOL without needing familiarity with commands, syntax or external scripts.

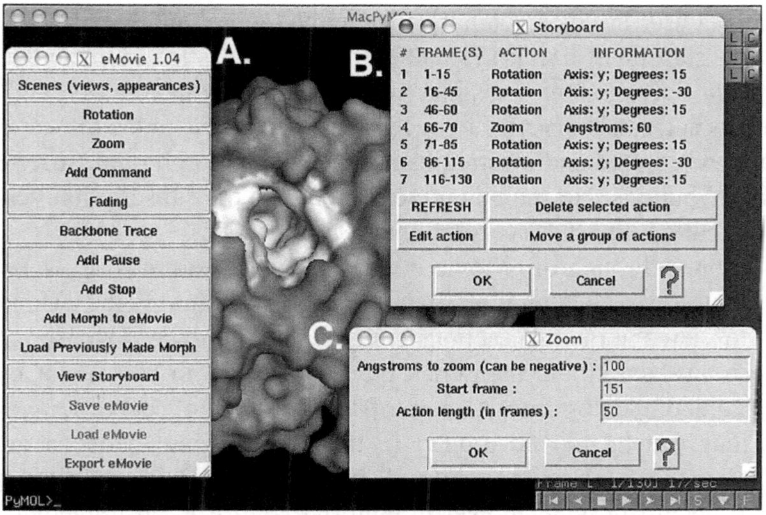

Figure 1. Snapshot of eMovie in action. In the background is the PyMOL GUI and the eMovie plug-in, with three open windows appears in the foreground. (A) The eMovie main menu bar with all of the buttons for inserting modular actions and viewing the storyboard (B) The storyboard window has been opened as a result of clicking "View Storyboard" on the eMovie menu bar. The storyboard displays the modular actions comprising the movie and invites the user to make changes. (C) The zoom window has been opened as a result of clicking "Zoom". A zooming action of 100 Å is being inserted starting at frame 151 and taking 50 frames to complete the action.

2.3. EMOVIE IN ACTION

eMovie (Fig. 1) runs on Macintosh OS X (the X11 version, i.e. PyMOLX11Hybrid),[1] UNIX/LINUX and Windows. The user is presented with several buttons for interactive movie production and instructive help buttons assist first-time users (Fig. 1A). eMovie expects the user to be familiar with PyMOL and its graphical user interface (GUI), which is mostly self-explanatory. In PyMOL, each movie is treated as an ordered sequence of frames (pictures), to which particular actions are assigned. eMovie enables the user to enter actions in a modular way, generally by defining the starting point, the duration in frames, and the type of event that is required (Fig. 1C). Each action appears on an editable storyboard (Fig. 1B).

2.4. CREATING A MOVIE

To create a movie using eMovie, the user first loads the "actors", or macromolecules, into PyMOL using the PyMOL GUI. Using the eMovie menu bar (Fig. 1A), the user chooses modular actions to add to different points in the movie. Modular actions include: scenes, rotations, zooms, fading, N->C rainbow colored backbone traces, and custom PyMOL commands. For example, a user inserts a zoom by clicking on 'zoom' and specifying a zoom of 100 Å starting at frame 151 and taking 50 frames to complete the action (Fig. 1C). Scenes store a distinct combination of view, colors, and representations and can be inserted to be recalled at any point in the movie. Morphs can also be created and inserted using the 'Make morph' and 'Insert morph to movie' buttons, but making a morph requires the subscription version of PyMOL called iPyMOL (version 0.99 recommended).

The breakthrough of eMovie is its storyboard feature (Fig. 1B). At any time the user can click on the "View Storyboard" button in the eMovie main menu to view the modular actions comprising the movie. The user can also click on any action and then click on 'Edit action' to edit it (for example change a zoom action from a zoom of 100 Å to one of 50 Å). Large groups of actions can be moved around in the movie using 'Move a group of actions', and actions can be removed entirely using 'Delete selected action'. The storyboard provides powerful editorial control in an immediately responsive and user-friendly manner. When the movie is finished, the finished

[1] Switching from the Macintosh version of PyMOL (MacPyMOL) to the X11 version of PyMOL (PyMOLX11Hybrid) is simply accomplished by renaming the filename of MacPyMOL as PyMOLX11Hybrid (switching the name back reverts from PyMOLX11Hybrid to MacPyMOL). The authors find it useful to keep a copy of both versions on the Mac.

version is saved using the "Save eMovie" button, which the user has been using throughout the movie-making process to periodically save his or her movie.

2.4.1. *Exporting the movie to a traditional file format*

When the animation is finished, there are two options for playing it in a program other than PyMOL. One option is to use the 'Export eMovie' button (Fig. 1A) to export the movie as an image sequence and to subsequently merge the sequence to a common movie format (e.g. .mov, .gif or .mpeg) using programs such as Mencoder (UNIX and Windows), Graphic Converter (Macintosh OS X), Movie Maker (Windows), VideoMach (Windows), Adobe Premiere (Windows), or Berkeley encode_mpeg (UNIX or LINUX). A second option (for Macintosh users) is to save the finished eMovie using the PyMOL 'File/Save Session' feature; this saved session .pse file can then be opened in MacPyMOL and exported as File/Save_movie/QuickTime. Other programs can be used to apply titles, voice-over, music, labels and end credits.

2.5. CONCLUDING REMARKS

The availability of a storyboard-based, molecular movie-making tool such as eMovie has, without a doubt, the potential to affect the way science is communicated, as evidenced by the quick adoption of and positive response to the predecessor of eMovie, the movie.py PyMOL plug-in (http://www.rubor.de/bioinf). By making the creation of molecular movies more fluid through an editable storyboard, eMovie enables more scientists to take advantage of this media. An unfamiliar user can prepare an illustrative molecular movie for a presentation or a website within hours or even minutes.

eMovie can be downloaded from http://www.weizmann.ac.il/ISPC/eMovie.html, and PyMOL can be downloaded from http://pymol.sourceforge.net.

Sections 2.1, 2.2, 2.3, 2.4.1, 2.5, and 4.1 are reprinted from *Trends in Biochemical Sciences*, **32**, "eMovie: a storyboard-based tool for making molecular movies", pp 199–204, Copyright (2007), with permission from Elsevier. For the full text including a more complete introduction to the field of molecular movie making and software that came before eMovie as well as a more detailed walkthrough of the eMovie workflow for creation of lengthy movies see the article in TiBS.

3. Proteopedia

3.1. INTRODUCTION

For all of the over 50,000 structures in the Protein Data Bank, rare is the non-structural biologist making use of the treasure of information. Because of the inherent complexity of a three-dimensional (3D) structure, structures can be hard to understand, especially for a non-structural biologist. As a result of this complexity, structural biology is still not in the mainstream of biology.

3D structures are most often communicated in two-dimensional (2D) media, namely scientific journals. In these journals 3D structures are compressed into 2D images, destroying the 3D information the crystallographer worked so hard to obtain. Molecular animations, such as those created with eMovie, are helpful in communicating 3D structures, but they are not the full solution because movies do not extend well to a large resource.

What's missing from structural biology is a way to communicate 3D structural information in a manner that is easy to understand. A resource with this quality would allow more effective communication between structural biologists and biologists, particularly by linking 3D structure to functional information in an intuitive manner.

Proteopedia is just this resource, and it has recently been made available on the web. In Proteopedia, descriptive text contains hyperlinks that, when clicked, change the orientation and representation of a 3D structure on the page in order to better explain a point made in the text. It is also a wiki-based web resource that allows the scientific community to easily contribute information via simple-to-use authoring tools. As a result, Proteopedia facilitates sharing and discussion among the scientific community and can help bring structural biologists and biologists to the same page.

Proteopedia is online at http://www.proteopedia.org, accessible via the popular operating systems and web browsers, and serving the scientific and educational community. There are already over 50,000 pages in the web resource, including pages on each of the entries in the PDB seeded with useful information and inviting an expert to contribute to them. Anyone from the scientific community can request an account to edit pages, but viewing pages does not require an account.

3.2. DESCRIPTIVE TEXT LINKED TO 3D VIEWS OF STRUCTURES

Perhaps the most powerful feature of Proteopedia is its capability to link descriptive text to 3D views of structures (a feature also implemented elegantly in the closed, proprietary viewer iSee[8]).

A user interested in Aricept®, the widely used drug for treatment of the symptoms of Alzheimer's, could search for "Aricept" in Proteopedia, and find the Proteopedia page for the PDB entry 1eve (Fig. 2A). The PDB entry 1eve is the crystal structure of Torpedo californica acetylcholinesterase (TcAChE) complexed with Aricept®. Upon loading the Proteopedia page entitled 1eve, the user is greeted with a slowly revolving 3D structure of Aricept®, also referred to as E2020, bound to the active site of TcAChE. The 3D structure is visualized using the molecular visualization applet Jmol.[9] While structures are not presented in true 3D, the rotation of the structure on the screen creates the illusion of 3D.[3]

The user begins reading the text on the page. She or he reads that the X-ray structure of the E2020-TcAChE complex shows that E2020 has a unique orientation along the active-site gorge, and clicks on the green text 'unique orientation'. Immediately the 3D structure responds by transitioning to a scene that showcases the unique orientation of E2020, Aricept®, along the active-site gorge (Fig. 2B). When the user reads that the orientation of E2020 extends from the anionic subsite (W84) of the active site, at the bottom, to the peripheral anionic site (near W279), at the top, he or she can click on the green links 'W84' and 'near W279' to see the respective residues labeled in the 3D scene (Fig. 2C,D). Reading that E2020 binds the active-site gorge tightly, but only through water intermediaries, the user clicks on the green link 'indirectly via solvent molecules' to view a scene illustrating the binding of E2020 in the active-site gorge (Fig. 2E).

We call these 3D views 'scenes' because they recall not just a particular view or orientation, but also a set of colors and representations and labels. At any time the user may zoom and rotate the 3D structure to explore it. The links can also be clicked in any order and still a smooth transition between 'scenes' results.

3.3. ADDING CONTENT AND EFFORTLESS CREATION OF 3D VIEWS

3.3.1. *How content is added*

Proteopedia is a 'wiki' web resource similar in spirit to Wikipedia.[10] This means that pages in Proteopedia are not static but can be changed and improved by its users. All a user needs to do to edit a page is click on the 'edit this page' tab at the top of the page and begin editing the text of the page or inserting 3D scenes. Text editing is simple and performed in the same way as in Wikipedia because both websites are based on the open-source MediaWiki[11] software for wiki-sites. The 3D scenes are fashioned, also in a simple manner, using the Proteopedia Scene Authoring Tools. In depth and up-to-date editing help and explanations are available at http://www.proteopedia.org/wiki/index.php/Help:Editing.

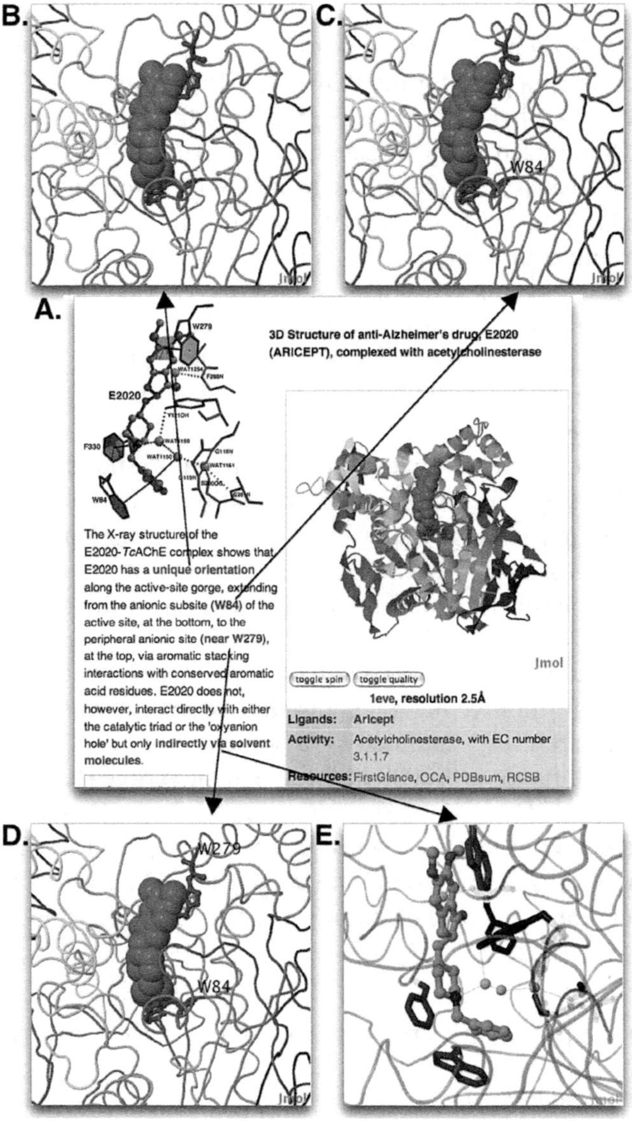

Figure 2. Proteopedia links descriptive text to 3D scenes. The arrows in this figure represent the changed Jmol visualization applet as a result of clicking on one of the green link (A) The page for the PDB entry 1eve containing the structure of AChE complexed with Aricept®. The molecule on the page is rotatable and zoomable and the green links change the appearance of the molecule when clicked. (A) The scene that loads when the user clicks on the green link 'unique orientation'. (B) W84 becomes labeled when the user clicks on the green link 'W84'. (C) W279 is labeled when the user clicks on the green link 'W279'. (D) Aricept® binds the active site of AChE tightly, but only through water intermediaries, this is show best when the user clicks on the green link 'indirectly via solvent molecules'.

3.3.2. *3D scenes in a snap*

Perhaps surprisingly, 3D scenes are incredibly easy to create in Proteopedia. 3D scenes are created using the Proteopedia Scene Authoring Tools (SAT) (Fig. 3), which are accessible from every Proteopedia page by clicking on that page's 'edit this page' tab. The concept behind the SAT is such: A user loads a 3D structure into the SAT using the 'load molecule' tab (Fig. 3A). Following that, the user manipulates the 3D scene to the desired orientation, coloring scheme, representations, and labeling that will appropriately showcase

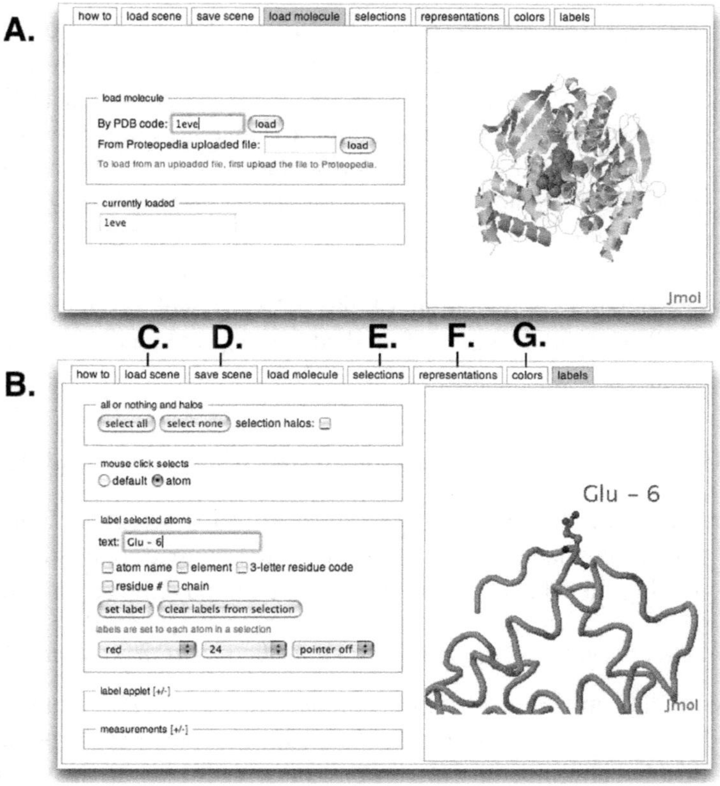

Figure 3. Two instances of the Proteopedia Scene Authoring Tools (SAT). (A) The SAT with the 'load molecule' tab active. The user has loaded the PDB file 1eve. (B) The SAT with the 'labels' tab active. The user has labeled a particular residue with the label 'Glu – 6' to highlight a mutation in hemoglobin that leads to the disease sickle-cell anemia. (C) The 'load scene' tab is used for loading existing scenes so they may be edited or used as a starting point for a new scene. (D) The 'save scene' tab allows the user to save a scene. (E) The 'selections' tab is used for choosing particular groups of atoms within the molecule. (F) The 'representations' tab is used for changing the representations of a particular group of atoms. (G) The 'colors' tab is used for changing the colors of a particular group of atoms, or the background color.

the intended point of the scene. When finished crafting the scene, the user saves the scene using the 'save scene' tab (Fig. 3D) and chooses a piece of text in the page to which to link the scene. That piece of text becomes a green link and recalls that scene when clicked.

In creating a scene, the user creates a still picture of sorts by saving a combination of colors, representations, labels, and viewpoint. When the scene is recalled by clicking the green link, the smooth transition that takes place is not a transition animation programmed by the user, but rather the transition animation is automatically handled by the software with each click of a green link.

The user sets the scene's orientation using the mouse. Colors, representations, and labels often need to be set differently for a different groups of atoms within the 3D structure. Thus, first the user uses the 'selections' tab (Fig. 3E) to choose a select group of atoms, and then uses the 'representations' tab (Fig. 3F) to change the group's representation, the 'colors' tab (Fig. 3G) to change the group's coloring scheme, and the 'labels' tab (Fig. 3B) to add labels to select atoms within the group. Previously created scenes may be recalled and edited using the 'load scenes' tab (Fig. 3C).

A more detailed explanation of the SAT as well as the text-editing features of Proteopedia can be found at www.proteopedia.org/index.php/Help:Editing.

3.4. THE CHALLENGES OF A WIKI FOR THE SCIENTIFIC COMMUNITY

In order for a wiki-based resource to succeed in the scientific world, some changes must be made to the traditional wiki creed that suggests that anyone viewing the site should be able to edit any page, even anonymously.

3.4.1. *Members only: accounts restricted to the scientific community*

In Proteopedia, only registered users may edit pages. This contrasts with Wikipedia's policy. To obtain a Proteopedia user account, a person must request one by clicking on 'log in / request account' at the upper right corner of the website. Account requests are approved only for members of the scientific community including scientists, educators, and students of science. In further contrast to Wikipedia, accounts are created in the users' real names to allow for credit to be given to worthwhile contributions, as well as to encourage users to take responsibility for their edits. For now credit is given in the form of an automatic listing of "contributors" at the bottom of each Proteopedia page, but this model may evolve in the future to better ascribe credit to deserving authors.

3.4.2. *Handling inaccuracies, the dangers of a wiki*

It is hoped that by limiting user accounts to the scientific community, inaccurate contributions will be kept to a minimum. Additional protection features in Proteopedia common to other wiki-sites include the ability to protect specific contested pages from editing except from a select group of expert stewards. Users knowledgeable on a particular protein or subject can elect to receive emails whenever the page is altered, thereby forming a sort of editorial board for the page. The user community, in its use and browsing of the web resource, will also be expected to correct inaccuracies. The 'recent changes' page in Proteopedia displays a list of the most recent changes made to the database so that vigilant users can patrol the new changes to judge their merit. A history is kept of every edit to a page, and the page can be reverted to any particular past version with the click of a mouse.

3.4.3. *Catering to educators and supplementary material*

In another departure from the traditional wiki model, Proteopedia offers each registered user a section of Proteopedia to 'own'. Whenever a new user account is confirmed, a user page for that user is created in Proteopedia. That page, as well as all sub-pages of that page are protected from editing and can only be edited by the user for which that particular user page is named. While editing of these pages by anyone other than their 'owner' is disallowed, copying useful content from 'owned' pages to other Proteopedia pages is allowed and encouraged where doing so will improve the content of Proteopedia's main pages.

The protected pages allow educators to take advantage of Proteopedia's 3D visualization features by posting lectures for projection on Proteopedia and knowing that they will be protected from edits. Similarly, structural biologists seeking a powerful way to communicate newly published research can post supplementary material on Proteopedia and have it protected from editing.

3.5. MORE THAN 50,000 PAGES AND COUNTING

While still in its infancy, Proteopedia already contains over 50,000 pages. Included among these pages are pages for each of the entries in the PDB. These PDB entry pages have been seeded with content to make them already useful as well as to provide a platform for users to add content. Each PDB entry page contains a 3D visualization of the structure with green links highlighting features occasionally defined in the PDB file such as the active site and ligands. Among other useful information, each PDB entry page contains the abstract from the structure's publication, inviting users to begin growing the page.

Proteopedia is not a copy of the PDB, however, in that the PDB entry pages are only part of the web resource. The next level up in the 'hierarchy' of pages would be the protein or molecule pages that have a more full description of the protein or molecule and also link to one or many relevant PDB entry pages. An example is the hemoglobin page, which gives a general overview of hemoglobin, as well as links to several PDB entry pages with various hemoglobin structures.

3.6. BEHIND THE SCENES: SOFTWARE

Proteopedia is based on an adapted version of the MediaWiki open source software for wiki-sites. Proteopedia integrates Jmol molecular visualization applets into the MediaWiki software via an adapted version of the Jmol MediaWiki Extension.[12]

3.7. DISCUSSION AND CONCLUSIONS

Proteopedia is growing daily with contributions. It is envisioned to be the first-stop resource for anyone interested in a particular protein or macro-molecule with a solved 3D structure. Ideally each structural biology lab would have it's own page on Proteopedia and the structural biology community as well as the scientific community at large would widely use the web resource. Only with increasing use over time can Proteopedia expect to satisfy its community of users and aid them in communicating structural biology to an ever-wider audience. The hope is that biologists will refer to Proteopedia to better understand the structure to function relationship of the proteins they are studying, and to better design experiments.

Proteopedia will continue to evolve and pursue future solutions for the intuitive presentation of 3D structural information. In its evolution, Proteopedia will continue to cater to its users by giving a high precedence to user-friendliness both in regards to users arriving to browse the resource and those arriving to contribute and add content.

ACKNOWLEDGEMENTS

EMOVIE

This study was supported by Autism Speaks, the Minerva Foundation, the Kalman and Ida Wolens Foundation, the Divadol Foundation, the Newman Foundation, a research grant from Mr. Erwin Pearl, the Bruce Rosen Foundation, the Kimmelman Center, the Israel Ministry of Science, Culture and Sport grant for the Israel Structural Proteomics Center (ISPC), and the European Commission Sixth Framework Research and Technological Development Programme 'SPINE2-COMPLEXES' Project under contract No. 031220. E.H. is grateful to the K. Kupcinet 2006 Summer School (Weizmann Institute) for a fellowship. J.L.S. is the Pickman Professor of Structural Biology. We thank Anat Katz for very helpful discussion on the eMovie manuscript and Michael George Lerner, Seth Harris, Laurence Pearl, Tserk Wassenaar and Lieven Buts for their contributions to the original movie.py, and Warren DeLano – the creator of PyMOL – for useful suggestions.

PROTEOPEDIA

This study was supported by Autism Speaks, the Nalvyco Foundation, the Jean and Julia Goldwurm Memorial Foundation, the Benoziyo Center for Neuroscience, the Divadol Foundation, the Neuman Foundation, a research grant from Mr. Erwin Pearl, the European Commission Sixth Framework Research and Technological Development Programme 'SPINE2-COMPLEXES' Project under contract No. 031220 and 'Teach-SG' Project, under contract number ISSG-CT-2007-037198. JLS is the Morton and Gladys Pickman Professor of Structural Biology. E.H. is grateful to the Karyn Kupcinet Program and the Feinberg Graduate School (Weizmann Institute of Science) for a fellowship. The authors are very grateful to the Jmol and MediaWiki development teams for their support and development of their respective software packages. We also greatly appreciate the useful discussions with Gideon Schreiber, Yigal Burstein, Harry Greenblatt, John Moult, Israel Silman, Eric Martz and Steven Brenner, as well as the generous use of content and images provided by Jane & David Richardson and David S. Goodsell.

References

1. Hodis, E., Schreiber, G., Rother, K. & Sussman, J. L., eMovie: A storyboard-based tool for making molecular movies, *TIBS* 32, 199–204 (2007).
2. Hodis, E., Prilusky, J., Martz, E., Silman, I., Moult, J. & Sussman, J. L., Proteopedia - a scientific 'wiki' bridging the rift between three-dimensional structure and function of biomacromolecules, *Genome Biol.* 9, R121 (2008).
3. Levinthal, C., Molecular model-building by computer, *Sci. Am.* 214, 42–52 (1966).
4. Richardson, D. C. & Richardson, J. S., The kinemage: A tool for scientific communication, *Protein Sci.* 1, 3–9 (1992).
5. Sayle, R. A. & Milner-White, E. J., RASMOL: biomolecular graphics for all, *TIBS* 20, 374–376 (1995).
6. Stampf, D. R., Felder, C. E. & Sussman, J. L., PDBBrowse - a graphics interface to the Brookhaven protein data bank, *Nature* 374, 572–574 (1995).

7. Peitsch, M. C., Stampf, D. R., Wells, T. N. C. & Sussman, J. L., The Swiss 3D-Image collection and Brookhaven protein data bank browser on the World-Wide Web, *TIBS* 20, 82–84 (1995).
8. Abagyan, R., Lee, W. H., Raush, E., Budagyan, L., Totrov, M., Sundstrom, M. & Marsden, B. D., Disseminating structural genomics data to the public: From a data dump to an animated story, *TIBS* 31, 76–78 (2006).
9. Jmol.
10. Wikipedia.
11. MediaWiki.
12. Vervelle, N. Jmol MediaWiki Extension.

STRUCTURAL STUDIES ON ACETYLCHOLINESTERASE AND PARAOXONASE DIRECTED TOWARDS DEVELOPMENT OF THERAPEUTIC BIOMOLECULES FOR THE TREATMENT OF DEGENERATIVE DISEASES AND PROTECTION AGAINST CHEMICAL THREAT AGENTS

JOEL L. SUSSMAN[1,2*], ISRAEL SILMAN[2,3]
[1]*Department of Structural Biology*
[2]*The Israel Structural Proteomics Center*
[3]*Neurobology Department, Weizmann Institute of Science, Rehovot, Israel*

Abstract. Acetylcholinesterase and paraoxonase are important targets for treatment of degenerative diseases, Alzheimer's disease and atherosclerosis, respectively, both of which impose major burdens on the health care systems in Western society. Acetylcholinesterase is the target of lethal nerve agents, and paraoxonase is under consideration as a bioscavenger for their detoxification. Both are thus the subject of research and development in the context of nerve agent toxicology. The crystal structures of the two enzymes are described, and structure/function relationships are discussed in the context of drug development and of development of means of protection against chemical threats.

Keywords: Alzheimer's disease, catalytic mechanism, crystal-forming variants, directed evolution, organophosphate, nerve agent

1. Introduction

In the following, we will discuss structure/function relationships for two enzymes which share the common feature that they are both associated with major neurodegenerative diseases, and are both relevant to prophylactic and

[*] To whom correspondence should be addressed. Joel L. Sussman, Department of Structural Biology, Weizmann Institute of Science, Rehovot 76100 ISRAEL; e-mail: joel.sussman@weizmann.ac.il

J.L. Sussman and P. Spadon (eds.), *From Molecules to Medicines*,

therapeutic efforts to overcome what is probably the most potent chemical threat, nerve agent intoxication.

The principal biological role of the synaptic enzyme, acetylcholinesterase (AChE), is to terminate synaptic transmission at cholinergic synapses by rapid hydrolysis of the neurotransmitter, acetylcholine (ACh).[1] In accordance with its biological role, it is an extremely rapid enzyme, operating at a rate at which diffusion control becomes rate-limiting.[2] Organophosphate (OP) nerve agents, and OP and carbamate insecticides, interact rapidly and irreversibly with the human and insect AChEs, respectively, with lethal consequences in both cases.[3,4]

Despite the lethal effects of nerve agents, cholinesterase inhibitors (ChEIs) have been used as drugs since the discovery of the pharmacological properties of the alkaloid, physostigmine, in the 1860s.[5] Interest in the medicinal chemistry of ChEIs intensified when the cholinergic hypothesis was put forward to explain the cognitive impairment characteristic of Alzheimer's disease (AD), the major form of senile dementia.[5,6] According to this hypothesis, cholinergic hypofunction in areas of the brain associated with cognition and memory occurs in AD, and can be symptomatically alleviated by administration of ChEIs. Indeed, all the first generation of drugs for the treatment of AD are ChEIs.[5,7]

Paraoxonase (PON), is a mammalian enzyme that, as its name implies, catalyzes the hydrolysis, and thereby the inactivation, of the insecticide, paraoxon, and of other OPs, such as the nerve agents, soman and sarin.[8,9] One form, PON1, is present in substantial amounts in the serum of humans and other mammals. In recent years it has become apparent that PON1 plays important roles in drug metabolism and in the prevention of the degenerative disease, atherosclerosis.[9,10] Indeed, knockout mice, totally lacking PON1, are highly susceptible both to atherosclerosis and to OP poisoning.[11] *In vitro* assays show that PON1 and the homologous PON3 inhibit lipid oxidation in low-density lipoprotein (LDL, 'bad cholesterol'), thus reducing levels of oxidized lipids that are involved in the initiation of atherosclerosis.[12,13] Because atherosclerosis is the underlying cause of 50% of mortality in Western societies, and OPs present an environmental risk as well as a terrorist threat, PONs have recently become the subject of intensive research.

The name paraoxonase is purely historical, since the PON family is a hydrolase family with one of the broadest specificities known, hydrolyzing OPs, neutral esters and lactones with varying degrees of efficacy. Recent evidence suggests that PON1 is a lactonase,[14] and homocysteine lactone is, indeed, a known risk factor for atherosclerosis.[15]

The solution of the crystal structures of *Torpedo californica* (*Tc*) AChE[16] and, more recently, of a variant of mammalian PON1 obtained by directed evolution,[17] provided the opportunity to obtain a detailed understanding of structure/function relationships in these two enzymes, which are of crucial importance for design of drugs for the treatment of two of the major degenerative diseases, and for developing approaches for coping with the major chemical threat posed by OP nerve agents.

2. Acetylcholinesterase

The 3D structure of *Tc*AChE displays a number of unexpected features (Fig. 1). Although, as already mentioned, AChE displays very high catalytic activity, especially for a serine hydrolase, its active site is deeply buried, being located almost 20 Å from the surface of the catalytic subunit, at the bottom of a long and narrow cavity. This cavity was named the active-site gorge or, since over 60% of its surface is lined by the rings of conserved aromatic residues, the aromatic gorge.[16,18] Despite the prediction that the catalytic 'anionic' site (CAS), which binds the quaternary moiety of ACh, would contain several negative charges,[19] in fact, only one negative charge is close to the catalytic site, that of E199, adjacent to the active-site serine, S200. Based both upon docking of ACh within the active site[16] and upon affinity labeling,[20] the principal interaction of the quaternary group appears to be a cation-π electron interaction,[21,22] with the indole ring of one of the conserved aromatic residues, W84. A second binding site for ACh at the entrance to the active-site gorge, named the peripheral anionic site (PAS), contains, as its principal residue, another conserved Trp residue, W279. The role of the PAS is to mediate substrate trapping,[23] the trapped ACh then proceeding down the gorge to the CAS. Further inspection reveals that *Tc*AChE possesses an unusually large dipole moment (>1,000 Debye), oriented along the axis of the active-site gorge, whose direction is such as to attract the positively charged substrate, ACh, down the gorge towards the active site.[24] Subsequent solution of the 3D structures of mouse AChE,[25] human AChE[26] and *Drosophila* AChE[27] showed that they all share these general features.

During the 18 years since the 3D structure of *Tc*AChE was solved, we have solved the structures of over 40 complexes and conjugates of a broad repertoire of inhibitors with this enzyme. In most cases, they were obtained by soaking the inhibitor into native crystals, but in a few cases co-crystallization was necessary, usually when crystal-packing constraints did not permit the ligand to occupy its binding site. The complexes formed with reversible inhibitors fall into three categories: (1) Ligands that bind at the active site at the bottom of the gorge; many of these are tertiary or quaternary amines that interact with the CAS; (2) Ligands that bind at the top of the gorge; these are known as 'peripheral site' ligands, since they interact with the PAS; (3) Ligands that span the gorge, containing aromatic and/or tertiary/quaternary amine moieties, linked by a spacer of a suitable length to permit them to span the CAS and the PAS.[7] The conjugates whose structures were solved were those of several potent OP nerve agents,[28,29] and of the anti-Alzheimer drug, rivastigmine (Exelon™), which is a carbamate.[30] All these compounds inactivate the enzyme irreversibly by attaching covalently to S200. In the following we will present the 3D structures of some prototypic complexes and conjugates, and discuss structure/function relationships.

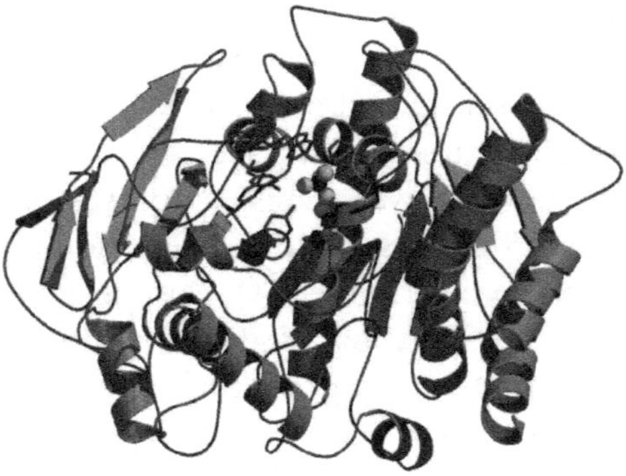

Figure 1. Ribbon diagram showing the 3D structure of *Tc*AChE. Conserved aromatic residues in the active-site gorge are represented as stick models, and ACh is shown as a space-filling model docked at the active site.

2.1. CONJUGATES

Conjugates with OP nerve agents were obtained by soaking the relevant OPs – sarin, soman and VX – into crystals of native *Tc*AChE.[28,29] The structures obtained initially, with sarin and soman, were those of 'aged' conjugates, i.e. of conjugates in which the bound OP moiety had undergone rapid and spontaneous dealkylation.[28,29] The structures of these conjugates revealed the exquisite fit of the OP moiety into the active site, which accounts for the potency of these highly toxic compounds. Thus, in addition to the covalent bond with S200, they make intimate interactions with the acyl pocket and with the oxyanion hole, and the nucleophilic oxygen generated by dealkyation makes a salt bridge with the imidazole of the catalytic histidine, H440.

In the case of VX, 'aging' is a very slow reaction, occurring on a time scale of weeks at 4°C. This permitted collection of structural data for the 'non-aged' conjugate immediately after soaking VX into the *Tc*AChE crystals, and for the non-aged conjugate a few weeks after soaking.[28] It was observed that insertion of the bulky OP moiety into the active site resulted in disruption of the catalytic triad. The H-bond between H440 and E327 was broken, and H440 formed a novel H-bond with E199, adjacent to the active-site serine. Upon 'aging', *viz.* removal of the ethyl group from the bound OP moiety, the H440-E327 H-bond was concomitantly restored (Fig. 2). Whereas OP conjugates with AChE are, in general, very stable, conjugates produced with carbamates usually turn over on a time scale of minutes.[31] The carbamyl conjugate of rivastigmine (Exelon™) with various AChEs, including the human enzyme, reactivates very slowly, and this may be partially responsible for the effectiveness of this drug.[30,32] Examination of the 3D structure of the conjugate of rivastigmine with *Tc*AChE revealed a structural basis for the slow reactivation observed. Just as in the conjugate with VX referred to above, the H440-E327 H-bond was disrupted; furthermore, the bulky aromatic [(dimethylamino)ethyl]phenol leaving group is retained in the active site.[30]

2.2. COMPLEXES

2.2.1. *Ligands binding at the CAS*

A diverse repertoire of ligands bind near the bottom of the active-site gorge. Most of them interact with W84, the principal residue contributing to the CAS, by either a π-cation interaction or a π-π stacking interaction.[7] Thus, edrophonium, a ChEI inhibitor that acts in the peripheral nervous system, and is used in the diagnosis of myasthenia gravis,[33] has a quaternary group that makes a π-cation interaction with W84. However its inhibitory capacity is significantly increased by formation of H-bonds to both S200 and H440 in the catalytic triad of the enzyme[34] (Fig. 3A). Tacrine (Cognex™), the first ChEI utilized for the treatment of Alzheimer's disease, though since discarded due its hepatoxicity,[35] also interacts with W84; but a second aromatic residue, F330, undergoes a conformational change, resulting in the formation of a sandwich in which the tacrine is stacked between the rings of the two aromatic residues[34] (Fig. 3B).

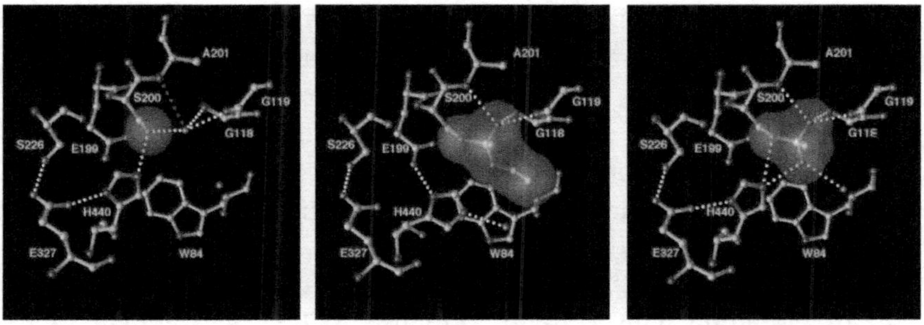

Figure 2. The active-site region in the crystal structures of *Tc*AChE prior to and after inhibition with the nerve agent, VX. (a) Native *Tc*AChE; (b) The 'non-aged' conjugate; (c) The 'aged' conjugate. The polypeptide chain is represented as a stick model, with H-bonds shown as dashed lines. The space-filling models correspond to S200Oγ in (a), to the 'non-aged' OP moiety in (b), and to the 'aged' OP moiety in (c).

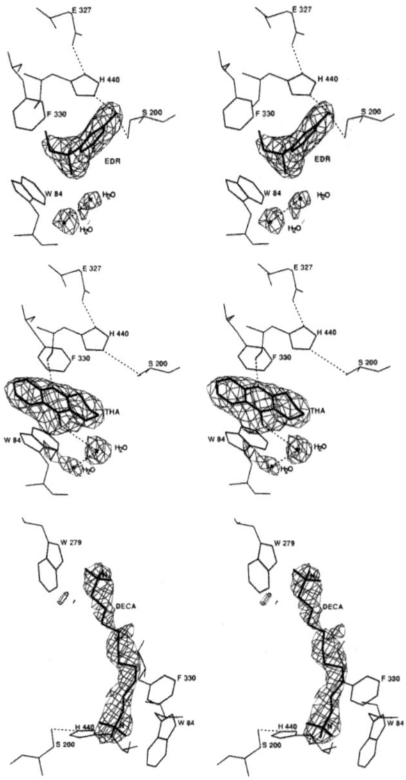

Figure 3. Electron density maps of complexes of *Tc*AChE with (a) edrophonium, (b) tacrine, and (c) decamethonium, showing the principal residues involved in interactions with the ligands. H-bonds are shown as broken lines.

2.2.2. *Ligands binding at the PAS*

Ligands binding specifically at the PAS are, in most cases, quaternary ligands that are too bulky to penetrate the active-site gorge, such as the Indian arrow poison, curare (*d*-tubocurarine),[36] and the fluorescent probe, propidium.[37] At the PAS, also, such ligands interact, via either π-cation or π-π interactions, with the conserved aromatic residues W279 and, to a lesser extent, Y70.[38] But the most striking example of a PAS-specific ligand is the three-fingered polypeptide, fasciculin, found in the venom of the black and green mambas that binds to AChEs from higher vertebrates with affinities in the range of 10^{-10}–10^{-12} M.[39] Solution of the crystal structures of complexes of fasciculin with *Tc*AChE,[40] mouse AChE[25] and human AChE,[26] revealed that the toxin completely covers the entrance to the active-site gorge (Fig. 4), displaying an unusually large contact area with the enzyme (>2,000 $Å^2$). Despite this large contact area, a crucial interaction between toxin and enzyme is that between the sulphur atom of M33 in the toxin and the indole group of W279 in *Tc*AChE (or the equivalent Trp residue in the mammalian AChEs). Mutation of this Trp residue results in a decrease of six orders of magnitude in the affinity of the toxin for the enzyme (Radic et al., 1994),[62] which explains its very low affinity for butyrylcholinesterase and for AChE of lower vertebrates, in which this residue is lacking.

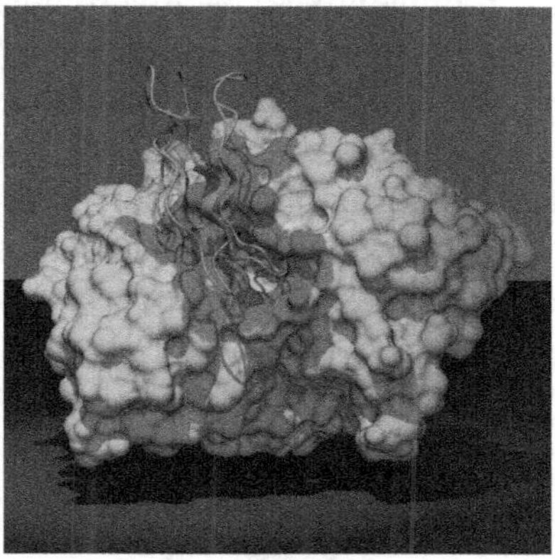

Figure 4. Crystal structure of the complex of fasciculin-II with human recombinant AChE. The molecular surface of the AChE molecule, with residues <4.0 Å from the fasciculin molecule rendered in dark grey, and residues within the active-site gorge rendered in white. The fasciculin molecule is displayed as a ribbon.

2.2.3. *Bifunctional gorge-spanning ligands binding at both the CAS and the PAS*

It has been shown in recent years that AChE can accelerate the assembly of the Aβ peptide to amyloid fibrils.[41,42] The inhibition of this process by propidium suggested that the PAS was responsible for this phenomenon. This raised the possibility that suitable gorge-spanning ligands could display a dual action as anti-Alzheimer drugs, serving as ChEIs at the CAS, and retarding assembly of Aβ to toxic amyloid fibrils at the PAS. This generated considerable interest in the medicinal chemistry of gorge-spanning ligands,[43,44] and in the development of bifunctional ligands that indeed displayed the capacity to inhibit both functions *in vitro*.[45] Furthermore, other categories of gorge-spanning ligands were developed in which the entity binding at the PAS served to inhibit another biological activity associated with neurodegeneration, such as monoamine oxidase,[46] as well as, most recently, gorge-spanning ligands with trifunctional capabilities.[47]

The prototypic gorge-spanning ligand is decamethonium, the ganglionic blocker in which two quaternary ammonium groups are separated by a decamethylene spacer,[48] thus permitting optimal interaction with both W84 and W279 of *Tc*AChE[34] (Fig. 3C). But in terms of utilizing the structural features of the active-site gorge of AChE for optimal drug-target interaction, the anti-Alzheimer drug E2020 (donepezil; Aricept™)[49] serves as an ideal paradigm. Figure 5 displays the detailed interactions of E2020 with *Tc*AChE as revealed by the crystal structure of the E2020/*Tc*AChE complex.[50] At the bottom of the gorge, the benzyl group of the drug stacks against W84, midway up the gorge the tertiary nitrogen of the piperidine ring makes a cation-π interaction with the phenyl ring of F330, and at the top of the gorge, the 5,6-dimethoxy-2,3-dihydroinden-1-one moiety stacks against W279. Interestingly, there are no direct H-bonds or salt bridges between the drug and the enzyme; but H-bonds are formed between the ligand and conserved water molecules within the active-site gorge[51]; these, in turn, form H-bonds with amino acid residues of the protein. Thus, in terms of drug design, the conserved water molecules can be considered as an integral part of the template formed by the surface of the gorge.

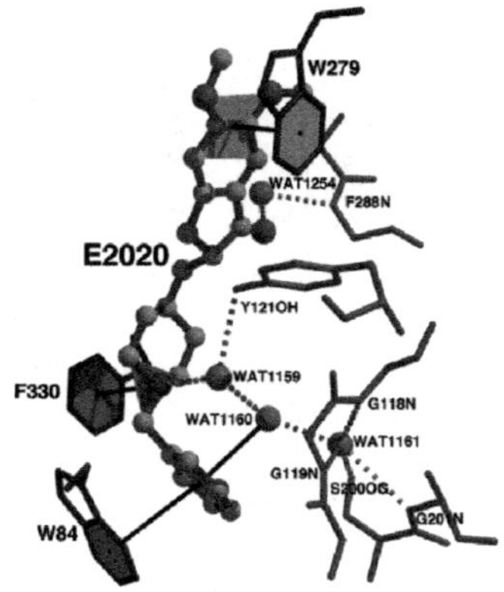

Figure 5. Binding modes of E2020 to *Tc*AChE. E2020 is displayed as a ball-and-stick model; water-mediated binding residues as light grey sticks, water molecules as light grey balls, "standard" H-bonds as heavy dashed lines, aromatic H-bonds, π cation and π–π stacking interactions as dark lines.

3. Paraoxonases

3.1. CRYSTALLIZATION AND STRUCTURE DETERMINATION

Human PON1 is rather unstable, and tends to aggregate in the absence of detergents.[52] Nor is it amenable to functional expression in bacteria or yeast, and thus to mutagenesis, library selection, and protein engineering. These factors led us to directly evolve PONs for bacterial expression and increased solubility.[53] Family shuffling of four PON1 genes (human, rabbit, mouse and rat) (Fig. 6), followed by screening for esterolytic activity, led to recombinant PON1 variants (rePON1s) that are expressed well in *E. coli*. These variants diverged from the wild-type (wt) rabbit PON1 by 14-32 amino acids contributed by the other three PON1 genes. The rePON1 variants exhibited enzymatic properties essentially identical to those of the wt PON1,[53] and similar biological activities in inhibiting LDL oxidation and mediating cholesterol efflux from macrophages.[54,55] Variants from the first round of evolution displayed a tendency to aggregate, and none could be crystallized. The second-generation variants, obtained by shuffling of the first-generation variants and screening for highest expression levels, did not aggregate, and one (G2E6) yielded stable well-diffracting crystals.

RePON1-G2E6 exhibits 91% homology to wt rabbit PON1 (Fig. 6), with the majority of the variations deriving from human, mouse, or rat weight PON1. Rabbit and human PON1s are also highly homologous in

```
PON1_RABBIT     MAKLTALTLLGLGLALFDGQKSSFQTRFNVHREVTPVELPNCNLVKGIDNGSEDLEILPN
G2E6            -----------M------R-----------------------------------------
G1C4            ----L----V--------------------------------------ET-A--------
G3C9            -----------------------------------------------V------------
G3H8            ------P-----------R-----------------------------------------
P1D6            ------------------R-----------------------------------------
P2D4            ------------------R-----------------------------------------

PON1_RABBIT     GLAFISSGLKYPGIMSFDPDKSGKILLMDLNEEDPVVLELGITGSTFDLSSFNPHGISTF
G2E6            --------------------------------KE-A-S--E-I-N-L-I-----------
G1C4            ------------------------------------------------------------
G3C9            --------------------------------------------N-L-I-----------
G3H8            ------------------------------------------------------------
P1D6            ------------------------------------------------------------
P2D4            ------------------------------------------------------------

PON1_RABBIT     TDEDNIVYLMVVNHPDSKSTVELFKFQEKEKSLLHLKTIRHKLLPSVNDIVAVGPEHFYA
G2E6            I-D--T---L-----G-S----V-----E-------------------------------
G1C4            -----T---L-------S----V-----E-------------------------------
G3C9            -----T---L-------S----V-----E-------------------------------
G3H8            -----T---L-------S----V---------------------------------S---
P1D6            -----T---L-------S----V-------R-------------------------S---
P2D4            ---------M------------------E-R-------------------------S---

PON1_RABBIT     TNDHYFIDPYLKSWEMHLGLAWSFVTYYSPNDVRVVAEGFDFANGINISPDGKYVYIAEL
G2E6            ------------------------------------------------------------
G1C4            ------------------------------------------------------------
G3C9            ------------------------------------------------------------
G3H8            ------A-----------------------------------------------------
P1D6            ------A-----------------------------------------------------
P2D4            ------------------------------------------------------------

PON1_RABBIT     LAHKIHVYEKHANWTLTPLKSLDFNTLVDNISVDPVTGDLWVGCHPNGMRIFYYDPKNPP

G2E6            -------------------RV-S-D---------------------------F--AE---
G1C4            ------------------------------------------------------------
G3C9            ------------------------D-----------------------------------
G3H8            ------------------------------------------------------------
P1D6            --------------------------------------------------------E---
P2D4            --------------------V-N-D---------------------------F--AE---

PON1_RABBIT     ASEVLRIQDILSKEPKVTVAYAENGTVLQGSTVAAVYKGKMLVGTVFHKALYCEL
G2E6            G-----------E------V--------------------L-I----------D-
G1C4            ------------ED--I--V----------T---S-----L-I----------D-
G3C9            G-----------E------V--------------------L-I------------
G3H8            G------------------V--------------------L-I----------D-
P1D6            G------------------V--------------------L-I----------D-
P2D4            G------------------V--------------S-----L-I----------D-
```

Figure 6. Sequence alignment of rabbit PON1 with six variants showing only residues differing from those in wild type rabbit PON1. G2E6 is the variant whose structure was solved, with its unique residues highlighted.

sequence (86 %) and function.[56] Sequence variations between rePON1-G2E6 and rabbit and human PON1 are in regions that do not affect their active sites.

The refined 2.2 Å crystal structure of rePON1 (R 18.5%; R_f 21.7%) contains one molecule per asymmetric unit. It was solved by single isomorphous replacement anomalous scattering (SIRAS) using a selenomethionine (SeMet) construct. The structure shows all residues except residues 1–15 at the N-terminus, and a surface loop (residues 72–79). Two calcium atoms, a phosphate ion, and 115 water molecules are also seen.

3.2. RABBIT PON1 VS. THE G2E6 VARIANT OBTAINED BY DIRECTED EVOLUTION

Figure 7 displays the crystal structure of the G2E6 variant, which shows 31 of the 32 residues that are mutated relative to the rabbit PON1 sequence (the mutation in residue 12 is in the N-terminal sequence that fails to display electron density). Twenty-two of these differences are on the surface of the molecule, and nine are in the core. The inner core mutations are from one hydrophobic residue to another. It has been suggested that the enhanced stability of G2E6 is due to three mutated residues, which are common to all the variants for which crystallization experiments were performed (Fig. 7). These are M341L and V343I, both hydrophobic core residues that pack

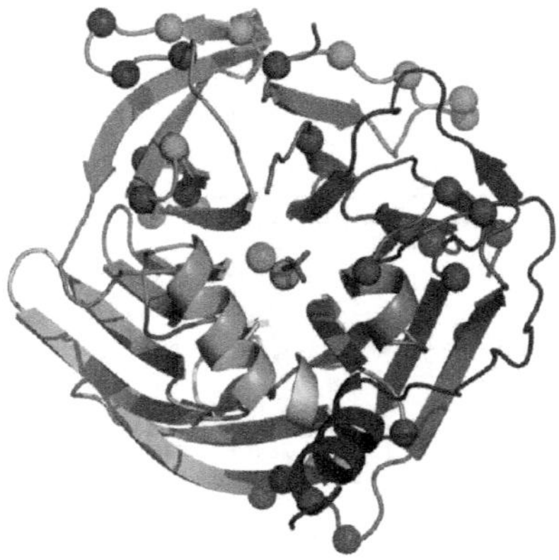

Figure 7. Ribbon diagram of the 6-bladed β-propeller structure of the rePON1 variant G6E2. The variant's mutated residues, relative to WT rabbit PON1, are shown as dark and light balls, light balls indicating unique mutations in G2E6 not shared by the five other variants of PON1 that failed to crystallize. The two Ca ions are shown as balls at the center of the structure, and the phosphate ion as a stick model. The two blades in the left-hand portion of the structure are mutation-free.

against each other, and the neighbouring A320V. It has been proposed[57] that these mutations increase the stability of the calcium-free apo-form of PON1.

Detergent-solubilized PON1 forms dimers and higher oligomers,[52] but there is only one molecule per asymmetric unit of the crystal, and very few contacts between symmetry-related molecules. It is possible that crystallization favours a monomeric form.

3.3. THE CATALYTIC MECHANISM

The upper of the two calcium ions seen in the crystal structure, Ca-1, lies at the bottom of the active-site cavity, together with a phosphate ion coming from the mother liquor.[17] One of this phosphate's oxygens is only 2.2 Å from Ca-1. This phosphate ion may be bound in a mode similar to the intermediates in the hydrolytic reactions catalyzed by PON, with the oxygen adjacent to Ca-1 mimicking the oxyanionic moiety of those intermediates stabilized by the positively-charged calcium ion. This type of "oxyanion hole" is seen in secreted phospholipase A2,[58] and has also been suggested for diisopropylfluorophosphatase,[59] whose overall structure closely resembles that of PON1. Two other phosphate oxygens may be mimicking the attacking hydroxyl ion and the oxygen of the alkoxy or phenoxy leaving groups of ester and lactone substrates.

To help elucidate PON1's mechanism, we determined its catalytic pH-rate profile. The pH-dependence observed may be ascribed to a histidine imidazole involved in a base-catalyzed, rate-determining step. In hydrolytic enzymes, histidine often serves as a base, deprotonating a water molecule, and thus generating the attacking hydroxide ion that produces hydrolysis.

A His-His dyad was identified near both Ca-1 and the phosphate ion (Fig. 8). We hypothesize that H115, the closer nitrogen of which is only 4.1 Å from Ca-1, acts as a general base to deprotonate a single water molecule, thus generating the attacking hydroxide, while H134 acts in a proton shuttle mechanism to increase H115's basicity. Interestingly, H115 adopts distorted dihedral angles - a phenomenon observed in catalytic residues of many enzymes. In support of the postulated mechanism we investigated the H115Q, H115A, and H115W mutations, all of which produce a dramatic decrease in both arylesterase and lactonase activity of PON1, and the H134Q mutation, which resulted in a milder, yet significant, decrease. Interestingly, the paraoxonase activity of the PON1 variant was not affected by any of these His mutations.[55]

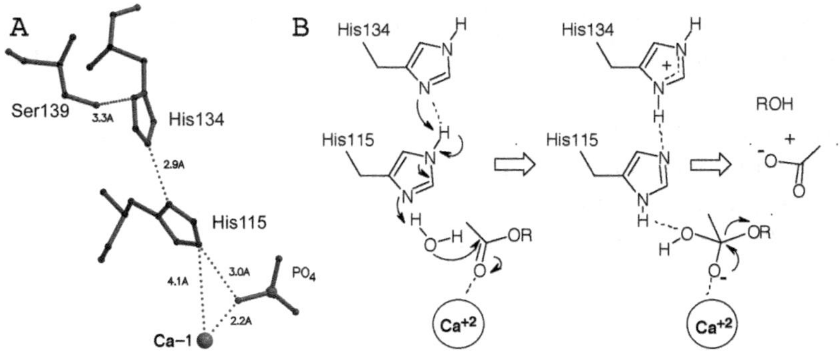

Figure 8. The postulated catalytic site and mechanism of PON1. (A) The catalytic site: the upper calcium atom (Ca-1), the phosphate ion at the bottom of the active site, and the postulated His-dyad; (B) Schematic representation of the proposed mechanism of action of PON1 on ester substrates such as phenyl and 2-naphthylacetate. The first step involves deprotonation of a water molecule by the His dyad to generate a hydroxide anion which attacks the ester carbonyl, producing an oxyanionic, tetrahedral intermediate. This intermediate breaks down (second step) to an acetate ion and either phenol or 2-naphthol.

3.4. DISCUSSION AND CONCLUSIONS

This study complements earlier examples of the application of directed evolution for protein crystallization.[60] Directed evolution was also applied to identify PON1's active site, and provided key insights as to how the substrate selectivity of the PON family members evolved in nature. Two thirds of the mutations in the G62E mutant are found on the surface of the molecule. The alignment of all the PON1 variants that were subjected to crystallization trials (Fig. 6) shows that the crystallized variant G2E6, which was crystallized successfully, has about twice as many mutations as those, which could not be crystallized. Thus, it bears 32 mutations vs. 13–19 mutations for the three variants that did not crystallize. Ten of these mutations are unique to the G2E6 variant. They are all located on the surface of PON1, yet they do not make contacts with symmetry-related molecules in the crystal. The present study shows the advantage of directed evolution in generating a range of variants that are very similar, of which only some may readily form crystals. The mutations in the G2E6 variant are spread throughout most of the molecule, except for two blades of the six-bladed β-propeller, blades 3 and 4, neither of which bears a mutation (Fig. 7). Detergent-solubilized PON1 forms dimers and higher oligomers,[52] although the crystal structure contains only one monomer in the asymmetric unit. It is possible that the mutation-free surface is the site of dimerization of the PON1, being in this respect similar to the highly soluble bacterial phosphotriesterase variants which contain only three point mutations, all of which are distal from the dimer surface[61]; i.e., in this latter case the dimer surface is also mutation-free.

The crystal structure unravels both the overall fold of the PON family and the details of PON1's structure. It permits postulation of the catalytic mechanism of the esterase and lactonase functions of PONs. The mutagenesis data[55] indicate that these two activities of PON1 are catalyzed by a H115-H134 dyad, while the paraoxonase activity is not affected by the dyad mutations and, hence, must be catalyzed by residues which have yet to be identified.

The directed evolution results demonstrate the remarkable evolvability of this enzyme family, and its impact on obtaining active, stable, and soluble protein variants that are amenable to crystallization and to subsequent 3-D structure determination.

ACKNOWLEDGEMENTS

We acknowledge financial support by Autism Speaks, the Nalvyco Foundation, the Bruce Rosen Foundation, a research grant from Mr. Erwin Pearl, the Jean and Jula Goldwurm Memorial Foundation, the Kimmelman Center for Biomolecular Structure and Assembly, and the Minerva Foundation to JLS, by the Benziyo Center for Neuroscience to IS, by the Israel Science Foundation to IS & JLS, and by Grant Number U54NS058183 from the National Institute of Neurological Disorders and Stroke and the Defense Threat Reduction Agency (DTRA) of the US Army. The structures were determined in collaboration with the Israel Structural Proteomics Center (ISPC), supported by the Israel Ministry of Science, Culture and Sport, the Divadol Foundation, the Neuman Foundation, the European Commission Sixth Framework Research and Technological Development Programme 'SPINE2-COMPLEXES' Project under contract No. 031220, and the 'Teach-SG' Project, under contract number ISSG-CT-2007-037198. JLS is the incumbent of the Morton and Gladys Pickman Chair of Structural Biology.

References

1. Rosenberry, T. L., Acetylcholinesterase, *Adv. Enzymol.* 43, 103–218 (1975).
2. Quinn, D. M., Acetylcholinesterase: enzyme structure, reaction dynamics, and virtual transition states, *Chem. Rev.* 87, 955–975 (1987).
3. Millard, C. B. & Broomfield, C. A., Anticholinesterases: medical applications of neurochemical principles, *J. Neurochem.* 64, 1909–1918 (1995).
4. Casida, J. E. & Quistad, G. B., Golden age of insecticide research: past, present, or future?, *Annu. Rev. Entomol.* 43, 1–16 (1998).
5. Giacobini, E. Cholinesterases Inhibitors: from the Calabar bean to Alzheimer therapy. In *Cholinesterases and Cholinesterase Inhibitors* (Giacobini, E., ed.), pp. 181–226. Martin Dunitz, London (2000).

6. Bartus, R. T., Dean, R. L. Jr., Beer, B. & Lippa, A. S., The cholinergic hypothesis of geriatric memory dysfunction, *Science* 217, 408–414 (1982).
7. Greenblatt, H. M., Dvir, H., Silman, I. & Sussman, J. L., Acetylcholinesterase: a multifaceted target for structure-based drug design of anticholinesterase agents for the treatment of Alzheimer's disease, *J. Mol. Neurosci.* 20, 369–384 (2003).
8. Aldridge, W. N., Serum esterases II. An enzyme hydrolysing diethyl *p*-nitrophenyl acetate (E600) and its identity with the A-esterase of mammalian sera, *Biochem. J.* 53, 117–124 (1953).
9. Draganov, D. I. & La Du, B. N., Pharmacogenetics of paraoxonases: a brief review, *Naunyn Schmiedebergs Arch. Pharmacol.* 369, 78–88 (2004).
10. Lusis, A. J., Atherosclerosis, *Nature* 407, 233–241 (2000).
11. Shih, D. M., Gu, L., Xia, Y. R., Navab, M., Li, W. F., Hama, S., Castellani, L. W., Furlong, C. E., Costa, L. G., Fogelman, A. M. & Lusis, A. J., Mice lacking serum paraoxonase are susceptible to organophosphate toxicity and atherosclerosis, *Nature* 394, 284–287 (1998).
12. Mackness, M. I., Arrol, S. & Durrington, P. N., Paraoxonase prevents accumulation of lipoperoxides in low-density lipoprotein, *FEBS Lett.* 286, 152–154 (1991).
13. Reddy, S. T., Wadleigh, D. J., Grijalva, V., Ng, C., Hama, S., Gangopadhyay, A., Shih, D. M., Lusis, A. J., Navab, M. & Fogelman, A. M., Human paraoxonase-3 is an HDL-associated enzyme with biological activity similar to paraoxonase-1 protein but is not regulated by oxidized lipids, *Arterioscler. Thromb. Vasc. Biol.* 21, 542–547 (2001).
14. Khersonsky, O. & Tawfik, D. S., Structure-reactivity studies of serum paraoxonase PON1 suggest that its native activity is lactonase, *Biochemistry* 44, 6371–6382 (2005).
15. Jakubowski, H., Calcium-dependent human serum homocysteine thiolactone hydrolase. A protective mechanism against protein N-homocysteinylation, *J. Biol. Chem.* 275, 3957–3962 (2000).
16. Sussman, J. L., Harel, M., Frolow, F., Oefner, C., Goldman, A., Toker, L. & Silman, I., Atomic structure of acetylcholinesterase from *Torpedo californica*: a prototypic acetylcholine-binding protein, *Science* 253, 872–879 (1991).
17. Harel, M., Aharoni, A., Gaidukov, L., Brumshtein, B., Khersonsky, O., Meged, R., Dvir, H., Ravelli, R. B., McCarthy, A., Toker, L., Silman, I., Sussman, J. L. & Tawfik, D. S., Structure and evolution of the serum paraoxonase family of detoxifying and anti-atherosclerotic enzymes, *Nat. Struct. Mol. Biol.* 11, 412–419 (2004).
18. Axelsen, P. H., Harel, M., Silman, I. & Sussman, J. L., Structure and dynamics of the active site gorge of acetylcholinesterase: synergistic use of molecular dynamics simulation and X-ray crystallography, *Protein Sci.* 3, 188–197 (1994).
19. Nolte, H. -J., Rosenberry, T. L. & Neumann, E., Effective charge on acetylcholinesterase active sites determined from the ionic strength dependence of association rate constants with cationic ligands, *Biochemistry* 19, 3705–3711 (1980).
20. Weise, C., Kreienkamp, H. -J., Raba, R., Pedak, A., Aaviksaar, A. & Hucho, F., Anionic subsites of the acetylcholinesterase from *Torpedo californica*: affinity labelling with the cationic reagent *N,N*-dimethyl-2-phenyl-aziridinium, *EMBO J.* 9, 3885–3888 (1990).
21. Dougherty, D. A. & Stauffer, D. A., Acetylcholine binding by a synthetic receptor: implications for biological recognition, *Science* 250, 1558–1560 (1990).
22. Dougherty, D. A., Cation-π interactions in chemistry and biology: a new view of benzene, phe, tyr, and trp, *Science* 271, 163–168 (1996).
23. Rosenberry, T. L., Johnson, J. L., Cusack, B., Thomas, J. L., Emani, S. & Venkatasubban, K. S., Interactions between the peripheral site and the acylation site in acetylcholinesterase, *Chem. Biol. Interactions* 157–158, 181–189 (2005).

24. Ripoll, D. R., Faerman, C. H., Axelsen, P. H., Silman, I. & Sussman, J. L., An electrostatic mechanism for substrate guidance down the aromatic gorge of acetylcholinesterase, *Proc. Natl. Acad. Sci. USA* 90, 5128–5132 (1993).
25. Bourne, Y., Taylor, P. & Marchot, P., Acetylcholinesterase inhibition by fasciculin: crystal structure of the complex, *Cell* 83, 503–512 (1995).
26. Kryger, G., Harel, M., Giles, K., Toker, L., Velan, B., Lazar, A., Kronman, C., Barak, D., Ariel, N., Shafferman, A., Silman, I. & Sussman, J. L., Structures of recombinant native and E202Q mutant human acetylcholinesterase complexed with the snake-venom toxin fasciculin-II, *Acta Cryst.* D56, 1385–1394 (2000).
27. Harel, M., Kryger, G., Rosenberry, T. L., Mallender, W. D., Lewis, T., Fletcher, R. J., Guss, J. M., Silman, I. & Sussman, J. L., Three-dimensional structures of *Drosophila melanogaster* acetylcholinesterase and of its complexes with two potent inhibitors, *Protein Sci.* 9, 1063–1072 (2000).
28. Millard, C. B., Koellner, G., Ordentlich, A., Shafferman, A., Silman, I. & Sussman, J. L., Reaction products of acetylcholinesterase and VX reveal a mobile histidine in the catalytic triad, *J. Am. Chem. Soc.* 121, 9883–9884 (1999).
29. Millard, C. B., Kryger, G., Ordentlich, A., Harel, M., Raves, M., Greenblatt, H. M., Segall, Y., Barak, D., Shafferman, A., Silman, I. & Sussman, J. L., Crystal structure of "aged" phosphylated *Torpedo* acetylcholinesterase: nerve agent reaction products at the atomic level, *Biochemistry* 38, 7032–7039 (1999).
30. Bar-On, P., Millard, C. B., Harel, M., Dvir, H., Enz, A., Sussman, J. L. & Silman, I., Kinetic and structural studies on the interaction of cholinesterases with the anti-Alzheimer drug rivastigmine, *Biochemistry* 41, 3555–3564 (2002).
31. Aldridge, W. N. & Reiner, E., *Enzyme Inhibitors as Substrates* (Elsevier, New York, 1975).
32. Weinstock, M., Razin, M., Chorev, M. & Enz, A., Pharmacological evaluation of phenyl-carbamates as CNS-selective acetylcholinesterase inhibitors, *J. Neural Transm. Suppl.* 43, 219–225 (1994).
33. Karatas, H., Nurlu, G. & Kansu, T., Is there still a role for edrophonium in diagnosing ocular myasthenia, *Eur. J. Neurosci.* 14, e4–5 (2007).
34. Harel, M., Schalk, I., Ehret-Sabatier, L., Bouet, F., Goeldner, M., Hirth, C., Axelsen, P., Silman, I. & Sussman, J. L., Quaternary ligand binding to aromatic residues in the active-site gorge of acetylcholinesterase, *Proc. Natl. Acad. Sci. USA* 90, 9031–9035 (1993).
35. Summers, W. K., Tachiki, K. H. & Kling, A., Tacrine in the treatment of Alzheimer's disease. A clinical update and recent pharmacologic studies, *Eur. Neurol.* 29(Suppl 3), 28–32 (1989).
36. Changeux, J. -P., Responses of acetylcholinesterase from *Torpedo marmorata* to salts and curarizing drugs, *Mol. Pharmacol.* 2, 369–392 (1966).
37. Taylor, P. & Lappi, S., Interaction of fluorescence probes with acetylcholinesterase. The site and specificity of propidium binding, *Biochemistry* 14, 1989–1997 (1975).
38. Bourne, Y., Taylor, P., Radic, Z. & Marchot, P., Structural insights into ligand interactions at the acetylcholinesterase peripheral anionic site, *EMBO J.* 22, 1–12 (2003).
39. Karlsson, E., Mbugua, P. M. & Rodriguez-Ithurralde, D., Fasciculins, anticholinesterase toxins from the venom of the green mamba *Dendroaspis angusticeps*, *J. Physiol. (Paris)* 79, 232–240 (1984).
40. Harel, M., Kleywegt, G. J., Ravelli, R. B. G., Silman, I. & Sussman, J. L., Crystal structure of an acetylcholinesterase-fasciculin complex: interaction of a three-fingered toxin from snake venom with its target, *Structure* 3, 1355–1366 (1995).

41. Inestrosa, N. C., Alvarez, A., Perez, C. A., Moreno, R. D., Vicente, M., Linker, C., Casanueva, O. I., Soto, C. & Garrido, J., Acetylcholinesterase accelerates assembly of amyloid-beta-peptides into Alzheimer's fibrils: possible role of the peripheral site of the enzyme, *Neuron* 16, 881–891 (1996).
42. Bartolini, M., Bertucci, C., Cavrini, V. & Andrisano, V., beta-Amyloid aggregation induced by human acetylcholinesterase: inhibition studies, *Biochem. Pharmacol.* 65, 407–416 (2003).
43. Du, D. M. & Carlier, P. R., Development of bivalent acetylcholinesterase inhibitors as potential therapeutic drugs for Alzheimer's disease, *Curr. Pharm. Des.* 10, 3141–3156 (2004).
44. Haviv, H., Wong, D. M., Silman, I. & Sussman, J. L., Bivalent ligands derived from Huperzine A as acetylcholinesterase inhibitors, *Curr. Top. Med. Chem.* 7, 375–387 (2007).
45. Belluti, F., Rampa, A., Piazzi, L., Bisi, A., Gobbi, S., Bartolini, M., Andrisano, V., Cavalli, A., Recanatini, M. & Valenti, P., Cholinesterase inhibitors: xanthostigmine derivatives blocking the acetylcholinesterase-induced beta-amyloid aggregation, *J. Med. Chem.* 48, 4444–4456 (2005).
46. Sterling, J., Herzig, Y., Goren, T., Finkelstein, N., Lerner, D., Goldenberg, W., Miskolczi, I., Molnar, S., Rantal, F., Tamas, T., Toth, G., Zagyva, A., Zekany, A., Finberg, J., Lavian, G., Gross, A., Friedman, R., Razin, M., Huang, W., Krais, B., Chorev, M., Youdim, M. B. & Weinstock, M., Novel dual inhibitors of AChE and MAO derived from hydroxy aminoindan and phenethylamine as potential treatment for Alzheimer's disease, *J. Med. Chem.* 45, 5260–5279 (2002).
47. Bolognesi, M. L., Cavalli, A., Valgimigli, L., Bartolini, M., Rosini, M., Andrisano, V., Recanatini, M. & Melchiorre, C., Multi-target-directed drug design strategy: from a dual binding site acetylcholinesterase inhibitor to a trifunctional compound against Alzheimer's disease, *J. Med. Chem.* 50, 6446–6449 (2007).
48. Bergmann, F., Wilson, I. B. & Nachmansohn, D., The inhibitory effect of stilbamidine, curare and related compounds and its relationship to the active groups of acetylcholine esterase. Action of stilbamidine upon nerve impulse conduction, *Biochim. Biophys. Acta* 6, 217–224 (1950).
49. Sugimoto, H., Ogura, H., Arai, Y., Limura, Y. & Yamanishi, Y., Research and development of donepezil hydrochloride, a new type of acetylcholinesterase inhibitor, *Jpn. J. Pharmacol.* 89, 7–20 (2002).
50. Kryger, G., Silman, I. & Sussman, J. L., Structure of acetylcholinesterase complexed with E2020 (Aricept®): implications for the design of new anti-Alzheimer drugs, *Structure* 7, 297–307 (1999).
51. Koellner, G., Kryger, G., Millard, C. B., Silman, I., Sussman, J. L. & Steiner, T., Active-site gorge and buried water molecules in crystal structures of acetylcholinesterase from *Torpedo californica*, *J. Mol. Biol.* 296, 713–735 (2000).
52. Josse, D., Ebel, C., Stroebel, D., Fontaine, A., Borges, F., Echalier, A., Baud, D., Renault, F., Le Maire, M., Chabrieres, E. & Masson, P., Oligomeric states of the detergent-solubilized human serum paraoxonase (PON1), *J. Biol. Chem.* 277, 33386–33397 (2002).
53. Aharoni, A., Gaidukov, L., Yagur, S., Toker, L., Silman, I. & Tawfik, D. S., Directed evolution of mammalian paraoxonases PON1 and PON3 for bacterial expression and catalytic specialization, *Proc. Natl. Acad. Sci. USA* 101, 482–487 (2004).
54. Rosenblat, M., Gaidukov, L., Khersonsky, O., Vaya, J., Oren, R., Tawfik, D. S. & Aviram, M., The catalytic histidine dyad of high density lipoprotein-associated serum paraoxonase-1 (PON1) is essential for PON1-mediated inhibition of low density lipoprotein oxidation and stimulation of macrophage cholesterol efflux, *J. Biol. Chem.* 281, 7657–7665 (2006).

55. Khersonsky, O. & Tawfik, D. S., The histidine 115-134 DYAD mediates the lactonase activity of mammalian serum paraoxonases, *J. Biol. Chem.* 281, 7649–7657 (2006).
56. Kuo, C. L. & La Du, B. N., Comparison of purified human and rabbit serum paraoxonases, *Drug Metab. Dispos.* 23, 935–944 (1995).
57. Roodveldt, C., Aharoni, A. & Tawfik, D. S., Directed evolution of proteins for heterologous expression and stability, *Curr. Opin. Struct. Biol.* 15, 50–56 (2005).
58. Sekar, K., Yu, B. Z., Rogers, J., Lutton, J., Liu, X., Chen, X., Tsai, M. D., Jain, M. K. & Sundaralingam, M., Phospholipase A2 engineering. Structural and functional roles of the highly conserved active site residue aspartate-99, *Biochemistry* 36, 3104–3114 (1997)
59. Scharff, E. I., Koepke, J., Fritzsch, G., Lucke, C. & Ruterjans, H., Crystal structure of diisopropylfluorophosphatase from *Loligo vulgaris*, *Structure (Camb)* 9, 493–502 (2001).
60. Waldo, G. S., Genetic screens and directed evolution for protein solubility, *Curr. Opin. Chem. Biol.* 7, 33–38 (2003).
61. Roodveldt, C. & Tawfik, D. S., Directed evolution of phosphotriesterase from *Pseudomonas diminuta* for heterologous expression in *Escherichia coli* results in stabilization of the metal-free state, *PEDS* 18, 51–58 (2005).
62 Radic, Z., Duran, R., Vellom, D. C., Li, Y., Chervenansky, C., & Taylor, P., Site of Fasciculin Interactions with acetylcholinesterase, *J. Biol. Chem.* 269, 11233–11239 (1994).

PROTEIN FUNCTION PREDICTION FROM STRUCTURE IN STRUCTURAL GENOMICS AND ITS CONTRIBUTION TO THE STUDY OF HEALTH AND DISEASE

JAMES D. WATSON*, JANET M. THORNTON
EMBL-European Bioinformatics Institute, Wellcome Trust Genome Campus, Hinxton, Cambridge CB10 1SD, UK

Abstract. The various structural genomics projects throughout the globe have developed high-throughput protein structure determination pipelines which have been responsible for the deposition of a vast number of protein structures. As a consequence of the need for rapid data release and their target selection strategy, these projects have deposited a large number of proteins with little or no functional information. As the experimental characterization of protein function is expensive and time consuming, the bioinformatics community was prompted to address the problem of protein function prediction from sequence and structure. Over the years many methods have been developed and show varying degrees of success. Here we will discuss the main types of approach, the problems faced and, with examples from the Midwest Center for Structural Genomics (MCSG), illustrate how these structures and the techniques developed can have a significant impact on the study of health and disease.

Keywords: Structural genomics, protein structure, function prediction, health and disease

1. Introduction

High-throughput structure determination projects, or structural genomics [1, 2], were designed to address the growing gap between the quantity of protein sequence data and that of protein structure data. The main focus of these projects was, through careful target selection, the identification and

* To whom correspondence should be addressed. James Watson, EMBL-European Bioinformatics Institute, Wellcome Trust Genome Campus, Hinxton, Cambridge CB10 1SD, UK; e-mail: watson@ebi.ac.uk

structural determination of novel proteins with previously unseen folds, which could help homology model much of the "sequence world". One consequence of the high-throughput target selection process is a large number of the structures solved have little or no functional information associated with them. This is in direct contrast to traditional structure determination projects where, in general, much is already known about the biochemical and biological function of a target protein; the structure serving to clarify the mechanistic details or the unique modifications that modulate function.

The experimental determination of a protein's function is expensive in terms of both time involved and resources required. In order to help overcome this problem a number of computational methods have been developed to predict protein function [3, 4], mostly involving the transference of function based on similarities between proteins. The first step is usually to identify sequence-based similarities and relationships to proteins of known function, but when these approaches provide little or no functional clues, the protein structure can often provide additional information. The structure-based methods range in scale from global fold and multimeric assembly comparisons, through localized clefts and pockets, down to highly specific arrangements of particular residues. Each level of comparison can provide similarities that indicate distant evolutionary relationships and functional adaptations not visible using current sequence methods.

Even with the current variety of methods available, the transference of function based on similarity is a difficult process. Examples can be found where proteins identical to one another have totally different functions solely due to their location in a cell or organism, as well as cases where apparently completely unrelated proteins are shown to have the same function [5]. It is clear that function prediction strategies taking into account as many different sources of information as possible are more likely to succeed, but it is also clear that in order to have confidence in the predictions there needs to be experimental validation. This has been the driving force behind recent developments in high-throughput experimental functional testing and ligand binding studies.

Overall the high-throughput structural biology projects have been successful at producing a large number of protein structures and automating almost every step in the structure determination process, but what has been the benefit with regards to studying health and disease? In the initial projects a number of the structural genomics centres included additional goals to select targets from particular pathogenic organisms or proteins with medicinal importance. A prime example is the Structural Genomics Consortium (SGC), a not-for-profit organization, which was set up with the specific goal to determine the three dimensional structures of proteins of medical relevance. On top of specific targets within each structural genomics centre, many of the structures solved primarily for their assumed novel fold have been found to have additional implications to the understanding of human proteins in health and disease. Thus the usefulness of the high-throughput technologies to these areas of research has been shown, and this has been supported by the recent funding of two centres for structural genomics of infectious diseases.

2. Structural genomics

Structural genomics projects were initially set up to determine, in a high-throughput manner, a large number of protein structures as rapidly and accurately as possible. In the United States the Protein Structure Initiative (PSI) was responsible for the funding of a number of large scale centres in a trial phase, of which a smaller number have been taken through to the full scale production phase (PSI-2) [6]. In addition to this, a number of other consortia have set up around the globe including Japan [7], Australia [8] and across Europe [9]. The numbers of structures solved as a result of these structural genomics initiatives has been enormous: between January 2000 and March 2007 just over 4,000 unique structural genomics PDB deposits were made, which represents almost 15% of all deposits from around the globe in the same timeframe. The majority of structures were solved using X-ray crystallography, although there were a significant number of NMR structures. A comprehensive analysis [10] of structural genomics targets from 11 consortia in 2005 indicated that approximately two thirds of all structural genomics domains in CATH [11] and SCOP [12] were novel, as compared to one fifth of non-structural genomics deposits solved in the same period.

Further investigation into the impact of these structures on the construction of reliable homology models indicated that, from the 316 structures used in the study, 9,000 non-redundant gene sequences could be modeled directly. In addition, one fifth of the models generated were based on sequence identities greater than 50%, so could potentially be useful for ligand docking studies and drug design. These studies and others suggests that the structural genomics approaches are highly successful at determining large numbers of high quality structures, that the structures are useful for modelling large proportions of the sequence databases, and most importantly a significant proportion of the models generated can be used for detailed small molecule and drug discovery studies.

3. Protein function prediction

3.1. DEFINITION OF FUNCTION

Protein function is not always well defined [13] as it can be described on a range of levels: from highly specific biochemical reactions, through pathways and processes up to the level of the whole organism and beyond. Even if a specific function is defined there are additional contextual problems faced when the function of a given protein varies depending on where and when it is produced. As an example [14] it is known that Rho and Protein Kinase D (PKD) are important signal transduction molecules in T-cells, but it was not known whether they act independently. An investigation revealed that Rho-dependent PKD regulation is observed when localized to the membrane, whereas cytosolic PKD acts independently of Rho.

In order to address this problem there have been numerous efforts to standardize functional descriptions and create controlled vocabularies. Initial attempts at standardization can be exemplified by the Enzyme Commission classification scheme, a four level hierarchy describing known enzymatic reactions. However, this scheme was only applicable to enzymes and had problems describing multifunctional proteins. More recently the Gene Ontology (GO) scheme [15] was devised, which standardizes biological terms through structured, controlled vocabularies that fall into three categories: "Cellular Component", "Biological Process" and "Molecular Function". GO has become the most widely used ontology in bioinformatics mainly due to the fact it is machine readable, which is reflected in the number of methods available that aim to predict function using GO terms [16–18].

3.2. FUNCTION PREDICTION METHODS: SEQUENCE-BASED

The starting point for all protein function prediction is an examination of the amino acid sequence. Generally speaking, when two proteins share over 40% sequence identity and they are homologues, then they are likely to share a similar function [19]. Below this point however the functional similarity falls away rapidly, but there are exceptions to these rules [20].

There are many sequence-based approaches that can be used to identify similarities between proteins. The most commonly used algorithms are BLAST [21] and FASTA [22], which are used to search protein sequence databases for similar sequences with known function. Increasingly powerful and sensitive profile-based searches can now identify ever more distant evolutionary relationships. Databases of protein families [23–25], structural and functional, can now be easily searched and often form the basis for many of the target selection and genome annotation projects. Other methods that utilize the protein sequence involve comparisons of genome structure, gene location analysis, protein-protein interaction data and regulatory gene network analyses.

3.3. FUNCTION PREDICTION METHODS: STRUCTURE-BASED

When the protein sequence provides little or few functional clues determining the structure can provide additional information and structural comparisons can identify more distant relationships than observed using sequence. The structure-based methods can generally be classified by the level at which they operate, ranging from large-scale features down to specific clusters of amino acids.

3.3.1. *Fold Matching*

The global fold of a protein is often conserved more than the sequence, which is reflected in the fact that although there is an infinite supply of protein sequences, the number of protein folds is expected to be in thousands. Care must be taken when comparing protein structures as similarity in protein fold without significant sequence similarity can be indicative of convergent evolution. In these cases the transfer of function is more difficult as the same fold can be used to perform a number of different functions.

Despite the complications in functional transfer, there are many different methods for comparing protein folds the most well known being DALI [26] (for a review of the various fold comparison servers see Novotney et al. [27]). Other algorithms include MSDfold [28] and CATHEDRAL [29] both of which use graph theory to match secondary structure elements, VAST [30] which uses vector alignment of secondary structures and CE [31] which uses combinatorial extension. On the whole the various methods tend to agree when compared using the same dataset, but disagreements are best resolved by using more than one server.

3.3.2. *Surface clefts and pockets*

The next level of detail focuses in on specific clefts or pockets on the protein surface. The active site of a protein is most commonly the largest cleft in the protein surface [32], although other regions can be involved in allosteric control of active sites. In the development of pharmaceuticals it is essential that other sites besides the active site are taken into account when looking at potential lead compounds as there could be additional effects if a compound binds to a different site.

Identification of pockets and clefts is most commonly achieved by fitting probe spheres (usually the size of a water molecule) to the surface of the protein and subsequently detecting gap regions [33]. The surfaces or pockets can then be compared with one another or to a database of known functional sites or ligand-binding sites. Examples of surface databases include the CASTp [34] surface library and the EF-site [35, 36] database. The comparison of protein surfaces has been extended in the MCSG to create the Global Protein Surface Survey (GPSS). Other types of algorithm use shape descriptors to characterize binding sites. One such method uses spherical harmonics to describe the shape of pockets and ligands [37], which simplifies the comparison to strings of integers making for rapid similarity searches.

Other methods go beyond physical shape comparisons and incorporate information of the physicochemical environment of the pocket. One such method, FEATURE [38], describes the local microenvironment of a site using not only detailed properties atomic or chemical groups but residue specific and secondary structure properties. Another method, SiteEngine [39], takes a different approach in that it represents residues as pseudo-centres of physicochemical properties which allows for rapid comparison and site matching using geometrical hashing. "SURF'S UP!" [40] also looks at physicochemical features on protein surfaces, but by using coarse grained surface features it can be used on comparative models, which is particularly relevant in structural genomics where one goal is improving homology model coverage.

3.3.3. *Residue templates*

Many enzymes and ligand binding sites are often characterized by highly conserved arrangements of a few key residues that are essential for the function of the protein. The classic example of this is the comparison of subtilisin and chymotrypsin, which have very similar active sites but completely different folds. In this case the serine-histidine-aspartate catalytic triad is conserved in three dimensional space but not in amino acid sequence. Detection of these similarities is difficult using sequence methods but is easily achieved using three dimensional template methods.

Initial methods using geometric hashing algorithms such as TESS are very effective for identifying similar clusters of residues but suffer from the need to preprocess the entire database of structures and store the data before a search can be performed. This algorithm has been superseded by the rapid JESS [41] algorithm, which can search an entire database of structures or templates on the fly. One such database of three dimensional templates is the Catalytic Site Atlas (CSA) [42] which was based on a manually curated dataset of known active sites but has been recently expanded using PSI-BLAST to identify additional homologues. Other types of template method include: Fuzzy Functional Forms (FFFs) [43], which uses distances between alpha carbons instead of 3D-atomic coordinates; SPASM/RIGOR [44], which uses C-alpha and side chain pseudo-atoms as its template; and PINTS

(Patterns In Non-homologous Tertiary Structures) [45], which detects the largest common three dimensional arrangement of residues between two structures.

Recently the idea of template scanning has been turned on its head with the "reverse template" approach of Laskowski et al. [46]. Rather than scanning an unknown structure for matches to a database of sites, the structure of interest is fragmented into lots of three residue templates and each template is scanned in turn against the Protein Data Bank (PDB). The automated selection of residues for the templates is based on the conservation score of the residues involved, the idea being that the most highly conserved residues will be conserved for a reason that is most likely to do with the protein function. Faced with many structural genomics targets with little or no sequence conservation, this automated approach may not always find the correct residues. It is for this reason that the method has been released as a separate server but with the added option to allow the user to select the residues of interest in the template generation process. This server, Tempura, is of most use when experimental evidence or the literature suggests particular amino acids are essential to a protein.

3.3.4. *Other methods*

There are a wide variety of other structure-based methods available for function prediction. Some methods are very specific to a particular functional class, one of the most popular being DNA-binding prediction methods. These methods range from structural motif comparisons (such as HTHquery [47] which recognizes DNA-binding helix-turn-helix motifs) to position specific scoring matrices, which use information on any given residue and its sequence neighbor to predict DNA-binding likelihood and can therefore be applied when no DNA binding homologues can be found.

Two methods using a database centric approach are MSDmotif and MSDsite [48]. Both databases can be used to identify similarities that can suggest functional relationships. MSDsite is focused on bound ligands and active sites of protein structures in the PDB, whereas MSDmotif concentrates on small structural motifs (such as Beta-turns, Asx-motifs and Nests [49]) that characterise active-sites.

3.4. COMBINING METHODS

It is evident that there are a great number of methods available for the prediction of function. Some methods are highly specific and focused on a single class of proteins, whereas others are more general and can identify multiple levels of function. Although each method can identify specific cases where it correctly predicts function, no single method is likely to be correct 100% of the time. It is therefore wise to take a more holistic approach and combine the analyses from as many sources as possible. As always, care must be taken when interpreting function predictions from any server, but when many different methods all point towards the same function the prediction is more likely to be correct.

A number of servers have been developed by various groups to combine the different sequence- and structure-based approaches. Some servers such as ProFunc [50] do not aim to provide a single predicted function. Rather, they act as a sort of meta-server where a user can submit a request to a number of useful methods in one go and get the results back in an easily understood format. Others, like ProKnow [51], also collect information from a combination of sequence and structural approaches but present an overall prediction (in the case of ProKnow the server uses Bayes' theorem to weight the significance of the all the results before presenting an overall function to the user in terms of GO annotation). A new server, ConFunc, also uses Gene Ontology terms but, in this case, to direct the function prediction process. The method produces a set of feature derived profiles from which a protein's function can be predicted.

3.5. EFFECTIVENESS OF METHODS

Considering the fact that the various structural genomics initiatives have been producing so many structures for a number of years, there have been surprisingly few studies into structure-based function prediction. An early review [52] of hypothetical proteins of known structure and their functional assignment provided some glimpses of the quality of functional assignments that can be made from structure. For this limited dataset detailed functional information was obtained for a quarter of them, half provided some functional information, but no functional information could be retrieved for the remaining quarter.

In 2003, Kim et al. [53]. analysed structures, solved at the Berkeley Center for Structural Genomics (BCSG) and with collaborators, where the structure provided functional or evolutionary clues. The examples demonstrated that some successful functional characterizations came from structural similarity undetectable in the sequence, others came from weak remote similarity combined with experimental testing, and some were as a direct result of the unexpected binding of ligands that turned out to be cofactors or substrates.

The largest analysis performed to date [54] examined the effectiveness of the ProFunc server for structure-based function prediction using structures solved by the Midwest Center for Structural Genomics (MCSG). Of all the structures solved by the MCSG, 93 proteins of known function were submitted to the ProFunc server and the top scoring matches from each method stored and backdated to the deposition date for each structure. The remaining top hit for each method was then manually compared with the known function for each protein and an assessment made as to whether the prediction was correct or not. The results from the study indicated that of the methods available in ProFunc, the fold recognition (MSDfold and DALI) and "reverse template" approaches were the most successful with about a 60% success rate.

4. Implications to the study of health and disease

Structural genomics has produced a large number of structures, improved automation of the structure determination process, prompted the integration of numerous databases and been key to the development of a number of standardization initiatives, but how has it helped in our understanding of health and disease?

4.1. FUNCTIONALLY SIGNIFICANT STRUCTURES

As discussed earlier, the various structural genomics centres were primarily set up to determine proteins of novel structure in the hope to increase the coverage of the protein "fold space". This has proven to be successful with all of the main centres depositing a number of new folds each year. However, it has been argued on numerous occasions [55, 56] that the solution of a protein fold just for the sake of it is of little practical use; rather it is the biological significance of structures that should be examined. Here the structural genomics initiatives have also been successful with the publication of numerous structures of significance to disease processes and infectious agents.

One such example is the AF0491 protein [57] from *A. fulgidus*, a homologue of the human Shwachman-Bodian-Diamond syndrome (SBDS) protein. SBDS is a rare autosomal recessive disorder caused by mutations in the SBDS gene on chromosome 7 and is characterized by abnormal pancreatic exocrine function, skeletal defects, and hematological dysfunction. The structure of the AF0491 archael homologue was determined as part of a structural genomics project. The structure revealed a three domain protein:

- C-terminal domain showed a commonly occurring fold making it difficult to infer function, however they are found in many RNA-binding and DNA-binding proteins.

- The central domain also adopts a common fold, the winged helix-turn-helix (wHTH). Although HTH domains [58] are widely used in DNA binding and have also been identified in RNA-binding proteins, this function is not likely for this domain as the surface of AF0491 does not have the expected general basic character for this function. Part of this domain may however be involved in protein-protein interactions.
- The N-terminal domain is a novel fold and is the location of most of the disease-linked mutations identified in SBDS patients. This new fold was found in a yeast protein, YHR087W, which prompted a search for shared function.

The identification of the yeast homologue provided additional experimental advantages not possible with the human protein. By studying the SBDS structural and sequence homologues (YHR087W and YLR022C respectively) both proteins were linked to RNA metabolism, supporting previous hypotheses based on bioinformatics studies.

4.2. PATHOGENIC ORGANISMS

In addition to structures with novel folds a number of the structures determined by structural genomics projects were selected from organisms which have significance to human health and disease. One such organism is *Pseudomonas aeruginosa*, an opportunistic human pathogen that has a natural resistance to many antibiotics and disinfectants. It causes infections throughout the body and is particularly significant to Cystic Fibrosis patients.

PA2721 [59] was structurally determined by the MCSG as part of their structural genomics project and found to be homodimeric. The structure was used to identify similar folds using 3DHit and DALI, both of which suggested similarity to the βαβββ superfamily, particularly fosfomycin resistance protein and Ni(II)-bound glyoxalase I. Additional structural analyses using ProFunc and the PvSOAR [60] server indicated the same region as possibly functionally relevant. Although the specific function of the protein in this case has not been confirmed, the pocket identified is similar to phosphatidylinositol specific phospholipase C, which suggests PA2721 could bind a phosphatidylinositol-like compound.

4.3. EXPERIMENTAL TESTING

A significant proportion of structures solved by the MCSG showed similarity to a variety of known enzymes and ligand binding sites. In these cases the bottleneck in the process became the experimental validation of the protein function. In order to improve the speed at which proteins could be experimentally tested, high-throughput enzyme assays were developed

[61]. The assays have relaxed substrate specificity and are designed to identify the subclass or sub-subclasses of enzymes from the following list of reactions: phosphatase, phosphodiesterase/nuclease, protease, esterase, dehydrogenase, and oxidase. Further biochemical characterization of proteins involves substrate screening with additional secondary screens using natural substrates.

5. Conclusions

Global structural genomics projects have been successful in the automation of much of the structure determination process and produced a vast number of structures with novel folds that can be used to homology model large numbers of protein sequences. A consequence of the high-throughput process and the requirement for rapid data release has resulted in a significant proportion of structures being deposited with little or no functional annotation. To address this many sequence- and structure-based methods have been developed to computationally predict the function of these new proteins. These methods have shown some success and though no one method provides all the answers a combined approach can provide confident predictions.

The structures solved by these high-throughput projects have not all been novel folds, many of them have been shown to be important to the study of human disease and others have been solved to help shed light on the biochemistry of common human pathogens. The use of high-throughput approaches to drug design [62, 63] and the study of infectious agents will become more important in the coming years with the funding of two centres for structural genomics of infectious diseases: the Centre for Structural Genomics of Infectious Diseases (CSGID) and the Seattle Structural Genomics Center for Infectious Disease (SSGCID). Both centres will be targeting proteins from a priority list from the National Institute of Allergy and Infectious Diseases and organisms causing emerging and re-emerging infectious diseases. The goal will be to assist the structure-based drug design of new therapeutics to combat infectious diseases.

References

1. Blundell T. L., Mizuguchi K. (2000) Structural genomics: an overview. *Prog. Biophys. Mol Biol.* 73:289–295.
2. Fox BG, Goulding C, Malkowski MG, Stewart L, Deacon A. (2008) Structural genomics: from genes to structures with valuable materials and many questions in between. *Nat Method.* 5(2):129–132.

3. Lee D, Redfern O, Orengo C. (2007) Predicting protein function from sequence and structure. *Nat Rev Mol Cell Biol.* 8(12):995–1005.
4. Watson JD, Laskowski RA, Thornton JM. (2005) Predicting protein function from sequence and structural data. *Curr Opin Struct Biol.* 15(3):275–284.
5. Whisstock JC, Lesk AM. (2003) Prediction of protein function from protein sequence and structure. *Q Rev Biophys* 36(3):307–340.
6. Service R. (2005) Structural biology. Structural genomics, round 2. *Science.* 307(5715):1554–1558.
7. Yokoyama S, Hirota H, Kigawa T, Yabuki T, Shirouzu M, Terada T, Ito Y, Matsuo Y, Kuroda Y, Nishimura Y, Kyogoku Y, Miki K, Masui R, Kuramitsu S. (2000) Structural genomics projects in Japan. *Nat Struct Biol.* 7 Suppl:943–945.
8. Puri M, Robin G, Cowieson N, Forwood JK, Listwan P, Hu SH, Guncar G, Huber T, Kellie S, Hume DA, Kobe B, Martin JL. (2006) Focusing in on structural genomics: the University of Queensland structural biology pipeline. *Biomol Eng.* 23(6):281–289.
9. Albeck S, Alzari P, Andreini C, Banci L, Berry IM, Bertini I, Cambillau C, Canard B, Carter L, Cohen SX, Diprose JM, Dym O, Esnouf RM, Felder C, Ferron F, Guillemot F, Hamer R, Ben Jelloul M, Laskowski RA, Laurent T, Longhi S, Lopez R, Luchinat C, Malet H, Mochel T, Morris RJ, Moulinier L, Oinn T, Pajon A, Peleg Y, Perrakis A, Poch O, Prilusky J, Rachedi A, Ripp R, Rosato A, Silman I, Stuart DI, Sussman JL, Thierry JC, Thompson JD, Thornton JM, Unger T, Vaughan B, Vranken W, Watson JD, Whamond G, Henrick K. (2006) SPINE bioinformatics and data-management aspects of high-throughput structural biology. *Acta Crystallogr D Biol Crystallogr.* 62(Pt 10):1184–1195.
10. Todd AE, Marsden RL, Thornton JM, Orengo CA. (2005) Progress of structural genomics initiatives: an analysis of solved target structures. *J Mol Biol.* 348(5):1235–1260.
11. Orengo CA, Michie AD, Jones S, Jones DT, Swindells MB, Thornton JM. (1997) CATH - a hierarchic classification of protein domain structures. *Structure* 5(8):1093–1108.
12. Lo Conte L, Ailey B, Hubbard T, Brenner S, Murzin A, Chothia C. (2000) SCOP: a structural classification of proteins database. *Nucleic Acids Res* 28(1):257–259.
13. Skolnick J, Fetrow J. (2000) From genes to protein structure and function: novel applications of computational approaches in the genomic era. *Trends Biotechnol.* 18(1):34–39.
14. Mullin MJ, Lightfoot K, Marklund U, Cantrell DA. (2006) Differential requirement for RhoA GTPase depending on the cellular localization of protein kinase D. *J Biol Chem.* 281(35):25089–25096.
15. Ashburner M, Ball CA, Blake JA, Botstein D, Butler H, Cherry JM, Davis AP, Dolinski K, Dwight SS, Eppig JT, Harris MA, Hill DP, Issel-Tarver L, Kasarskis A, Lewis S, Matese JC, Richardson JE, Ringwald M, Rubin GM, Sherlock G. (2000) Gene ontology: tool for the unification of biology. The Gene Ontology Consortium. *Nat Genet* 25(1):25–29.
16. Pazos F, Sternberg MJ. (2004). Automated prediction of protein function and detection of functional sites from structure. *Proc Natl Acad Sci USA.* 101: 14754–14759.
17. Guo X, Shriver CD, Hu H, Liebman MN. (2005) Analysis of metabolic and regulatory pathways through Gene Ontology-derived semantic similarity measures. AMIA. *Annu. Symp. Proc.* 972.
18. Lee V, Camon E, Dimmer E, Barrell D, Apweiler, R. (2005) Who tangos with GOA?-use of Gene Ontology Annotation (GOA) for biological interpretation of '-omics' data and for validation of automatic annotation tools. *In Silico. Biol.* 5, 5–8.
19. Todd A, Orengo C, Thornton JM. (2001) Evolution of function in protein superfamilies, from a structural perspective. *J Mol Biol.* 307(4):1113–1143.
20. Rost B. (2002) Enzyme function less conserved than anticipated. *J Mol Biol.* 318(2):595–608.

21. Altschul SF, Madden TL, Schaffer AA, Zhang J, Zhang Z, Miller W, Lipman DJ. (1997) Gapped BLAST and PSI-BLAST: a new generation of protein database search programs. *Nucleic Acids Res.* 25(17):3389–3402.
22. Pearson WR. (1991) Searching protein sequence libraries: comparison of the sensitivity and selectivity of the Smith-Waterman and FASTA algorithms. *Genomics.* 11(3):635–650.
23. Finn RD, Mistry J, Schuster-Böckler B, Griffiths-Jones S, Hollich V, Lassmann T, Moxon S, Marshall M, Khanna A, Durbin R, Eddy SR, Sonnhammer ELL and Bateman A. (2006) Pfam: clans, web tools and services. *Nucleic Acids Res.* 34(Database Issue): D247–D251.
24. Mulder NJ, Apweiler R, Attwood TK, Bairoch A, Bateman A, Binns D, Bork P, Buillard V, Cerutti L, Copley R, Courcelle E, Das U, Daugherty L, Dibley M, Finn R, Fleischmann W, Gough J, Haft D, Hulo N, Hunter S, Kahn D, Kanapin A, Kejariwal A, Labarga A, Langendijk-Genevaux PS, Lonsdale D, Lopez R, Letunic I, Madera M, Maslen J, McAnulla C, McDowall J, Mistry J, Mitchell A, Nikolskaya AN, Orchard S, Orengo C, Petryszak R, Selengut JD, Sigrist CJA, Thomas PD, Valentin F, Wilson D, Wu CH, Yeats C. (2007) New developments in the InterPro database. *Nucleic Acids Res* 35(Database Issue): D224–D228.
25. Gough J, Chothia C. (2002) SUPERFAMILY: HMMs representing all proteins of known structure. SCOP sequence searches, alignments and genome assignments. *Nucleic Acids Res.* 30(1):268–272.
26. Holm L, Sander C. (1995) Dali: a network tool for protein structure comparison. *Trends Biochem Sci.* 20(11):478–480.
27. Novotny M, Madsen D, Kleywegt GJ. (2004) Evaluation of protein fold comparison servers. *Proteins.* 54(2):260–270.
28. Krissinel E, Henrick K. (2004) Secondary-structure matching (SSM), a new tool for fast protein structure alignment in three dimensions. *Acta Crystallogr D Biol Crystallogr.* 60(Pt 12 Pt 1):2256–2268.
29. Redfern OC, Harrison A, Dallman T, Pearl FM, Orengo CA. (2007) CATHEDRAL: a fast and effective algorithm to predict folds and domain boundaries from multidomain protein structures. *PLoS Comput Biol.* 3(11):e232.
30. Madej T, Gibrat JF, Bryant SH. (1995) Threading a database of protein cores. *Proteins* 1995 Nov; 23(3):356–3690.
31. Shindyalov IN, Bourne PE. (1998) Protein structure alignment by incremental combinatorial extension (CE) of the optimal path. *Protein Eng.* 11(9):739–747.
32. Laskowski RA, Luscombe N, Swindells M, Thornton JM. (1996) Protein clefts in molecular recognition and function. *Protein Sci.* 5(12):2438–2452.
33. Laskowski RA. (1995) SURFNET: a program for visualizing molecular surfaces, cavities, and intermolecular interactions. *J Mol Graph.* 13(5):323–328.
34. Dundas J, Ouyang Z, Tseng J, Binkowski A, Turpaz Y, Liang J. (2006) CASTp: computed atlas of surface topography of proteins with structural and topographical mapping of functionally annotated residues. *Nucleic Acids Res.* 34(Web Server issue):W116–8.
35. Kinoshita K, Nakamura H. (2004) eF-site and PDBjViewer: database and viewer for protein functional sites. *Bioinformatics.* 20(8):1329–1330.
36. Standley DM, Kinjo AR, Kinoshita K, Nakamura H. (2008) Protein structure databases with new web services for structural biology and biomedical research. *Brief Bioinform* 9:276–85.
37. Morris RJ, Najmanovich RJ, Kahraman A, Thornton JM. (2005) Real spherical harmonic expansion coefficients as 3D shape descriptors for protein binding pocket and ligand comparisons. *Bioinformatics.* 21(10):2347–2355.

38. Wei L, Altman RB. (2003) Recognizing complex, asymmetric functional sites in protein structures using a Bayesian scoring function. *J Bioinform Comput Biol.* 1(1):119–138.
39. Shulman-Peleg A, Nussinov R, Wolfson HJ. (2004) Recognition of functional sites in protein structures. *J Mol Biol.* 339(3):607–33.
40. Sasin JM, Godzik A, Bujnicki JM. (2007) SURF'S UP! - protein classification by surface comparisons. *J Biosci.* 32(1):97–100.
41. Barker JA, Thornton JM. (2003) An algorithm for constraint-based structural template matching: application to 3D templates with statistical analysis. *Bioinformatics* 19(13): 1644–1649.
42. Porter CT, Bartlett GJ, Thornton JM. (2004) The Catalytic Site Atlas: a resource of catalytic sites and residues identified in enzymes using structural data. *Nucleic Acids Res.* 32(Database issue):D129–33.
43. Fetrow JS, Skolnick J. (1998) Method for prediction of protein function from sequence using the sequence-to-structure-to-function paradigm with application to glutaredoxins/ thioredoxins and T1 ribonucleases. *J Mol Biol.* 281(5):949–968.
44. Kleywegt GJ. (1999) Recognition of spatial motifs in protein structures. *J Mol Biol.* 285:1887–1897.
45. Stark A, Shkumatov A, Russell RB. (2004) Finding functional sites in structural genomics proteins. *Structure* 12:1405–1412.
46. Laskowski RA, Watson JD, Thornton JM. (2005) Protein function prediction using local 3D templates. *J Mol Biol* 351(3):614–626.
47. Ferrer-Costa C, Shanahan HP, Jones S, Thornton JM. (2005) HTHquery: a method for detecting DNA-binding proteins with a helix-turn-helix structural motif. *Bioinformatics.* 21(18):3679–3680.
48. Golovin A, Dimitropoulos D, Oldfield T, Rachedi A and Henrick K. (2005) MSDsite: A Database Search and Retrieval System for the Analysis and Viewing of Bound Ligands and Active Sites. PROTEINS: *Structure, Function Bioinformat.* 58(1):190–199.
49. Watson JD, Milner-White EJ. (2002) A novel main-chain anion-binding site in proteins: the nest. A particular combination of phi,psi values in successive residues gives rise to anion-binding sites that occur commonly and are found often at functionally important regions. *J Mol Biol.* 315(2):171–182.
50. Laskowski RA, Watson JD, Thornton JM. (2005) ProFunc: a server for predicting protein function from 3D structure. *Nucleic Acids Res.* 33(Web Server issue):W89–W93.
51. Pal D, Eisenberg D. (2005) Inference of protein function from protein structure. *Structure* 13(1):121–130.
52. Teichmann SA, Murzin AG, Chothia C. (2001) Determination of protein function, evolution and interactions by structural genomics. *Curr Opin Struct Biol.* 11(3):354–363.
53. Zhang C, Kim S-H. (2003) Overview of structural genomics: from structure to function. *Curr Opin Chem Biol.* 7:28–32.
54. Watson JD, Sanderson S, Ezersky A, Savchenko A, Edwards A, Orengo C, Joachimiak A, Laskowski RA, Thornton JM. (2007) Towards fully automated structure-based function prediction in structural genomics: a case study. *J Mol Biol.* 367:1511–1522.
55. Cyranoski D. (2006) 'Big science' protein project under fire. Nature 443: 382.
56. Petsko GA. (2007) An idea whose time has gone. *Genome Biol.* 8(6):107.
57. Savchenko A, Krogan N, Cort JR, Evdokimova E, Lew JM, Yee AA, Sánchez-Pulido L, Andrade MA, Bochkarev A, Watson JD, Kennedy MA, Greenblatt J, Hughes T, Arrowsmith CH, Rommens JM and Edwards AM. (2005) The Shwachman-Bodian-Diamond Syndrome Protein Family Is Involved in RNA Metabolism. *J Biol Chem.* 280(19):19213–19220.

58. Aravind L, Anantharaman V, Balaji S, Babu MM, Iyer LM. (2005) The many faces of the helix-turn-helix domain: transcription regulation and beyond. *FEMS Microbiol Rev* 29(2):231–262.
59. Nocek B, Cuff M, Evdokimova E, Edwards A, Joachimiak A, Savchenko A. (2006) 1.6 A crystal structure of a PA2721 protein from pseudomonas aeruginosa - a potential drug-resistance protein. *Proteins.* 63:1102–1105.
60. Binkowski TA, Freeman P, Liang J. (2004) pvSOAR: detecting similar surface patterns of pocket and void surfaces of amino acid residues on proteins. *Nucleic Acids Res.* 32:W555–W558.
61. Kuznetsova E, Proudfoot M, Sanders SA, Reinking J, Savchenko A, Arrowsmith CH, Edwards AM, Yakunin AF. (2005) Enzyme genomics: Application of general enzymatic screens to discover new enzymes. *FEMS Microbiol Rev* 29(2):263–279.
62. Allali-Hassani A, Pan PW, Dombrovski L, Najmanovich R, Tempel W, Dong A, Loppnau P, Martin F, Thornton J, Edwards AM, Bochkarev A, Plotnikov AN, Vedadi M, Arrowsmith CH. (2007) Structural and Chemical Profiling of the Human Cytosolic Sulfotransferases. *PLoS Biol.* 5(5): e97.
63. Weigelt J, McBroom-Cerajewski LD, Schapira M, Zhao Y, Arrowmsmith CH. (2008) Structural genomics and drug discovery: all in the family. *Curr Opin Chem Biol.* 12(1): 32–39.

CRYSTAL STRUCTURES OF THE β_2-ADRENERGIC RECEPTOR

WILLIAM I. WEIS[*,1], DANIEL M. ROSENBAUM, SØREN G.F. RASMUSSEN, HEE-JUNG CHOI[1], FOON SUN THIAN, TONG SUN KOBILKA, XIAO-JIE YAO, PETER W. DAY, CHARLES PARNOT, JUAN J. FUNG, VENKATA R.P. RATNALA, BRIAN K. KOBILKA
Departments of Molecular & Cellular Physiology and [1]Structural Biology, Stanford University School of Medicine, 279 Campus Drive West, Stanford, CA 94305 USA

VADIM CHEREZOV, MICHAEL A. HANSON, PETER KUHN[2], RAYMOND C. STEVENS
Department of Molecular Biology and [2]Department of Cell Biology, The Scripps Research Institute, La Jolla, CA, 92037 USA

PATRICIA C. EDWARDS, GEBHARD F.X. SCHERTLER
MRC Laboratory of Molecular Biology, Cambridge, UK

MANFRED BURGHAMMER
European Synchrotron Radiation Facility, Grenoble, France

RUSLAN SANISHVILI, ROBERT F. FISCHETTI
Biosciences Division, Argonne National Laboratory, IL, USA

ASNA MASOOD, DANIEL K. ROHRER
Medarex, Inc., 521 Cottonwood Drive, Milpitas, CA 95035, USA

Abstract. G protein coupled receptors (GPCRs) constitute the largest family of membrane proteins in the human genome, and are responsible for the majority of signal transduction events involving hormones and neurotransmitters across the cell membrane. GPCRs that bind to diffusible ligands have low natural abundance, are relatively unstable in detergents, and display basal G protein activation even in the absence of ligands. To overcome these problems two approaches were taken to obtain crystal structures of the

[*] To whom correspondence should be addressed. William Weis, Departments of Structural Biology and Molecular & Cellular Physiology; e-mail: bill.weis@stanford.edu

J.L. Sussman and P. Spadon (eds.), *From Molecules to Medicines*,

β_2-adrenergic receptor (β_2AR), a well-characterized GPCR that binds catecholamine hormones. The receptor was bound to the partial inverse agonist carazolol and co-crystallized with a Fab made to a three-dimensional epitope formed by the third intracellular loop (ICL3), or by replacement of ICL3 with T4 lysozyme. Small crystals were obtained in lipid bicelles (β_2AR-Fab) or lipidic cubic phase (β_2AR-T4 lysozyme), and diffraction data were obtained using microfocus technology. The structures provide insights into the basal activity of the receptor, the structural features that enable binding of diffusible ligands, and the coupling between ligand binding and G-protein activation.

1. Introduction

The G-protein coupled receptors (GPCRs) are the largest family of membrane receptors in the human genome, and are responsible for the majority of responses to hormones and neurotransmitters.[1,2] GPCRs share a common architecture consisting of seven transmembrane (TM)-spanning helices, with an extracellular N-terminus and intracellular C-terminus (Fig. 1). Interaction of a GPCR with an activating ligand triggers conformational changes that catalyze the exchange of GDP for GTP in the α-subunit of a cognate heterotrimeric G protein. The GTP-bound G protein dissociates from the receptor to mediate downstream signaling events. GPCRs are the targets of a significant fraction of pharmaceuticals currently in use, but structure-based drug design efforts have been hindered by the lack of three-dimensional structural information.

The β_2-adrenergic receptor (β_2AR) is a very well-studied GPCR that is activated by catecholamine ligands, and has important roles in cardiovascular and pulmonary physiology.[2,3] It is a member of the largest subclass of GPCRs, the so-called type A receptors, that includes adrenergic receptors, the visual pigment rhodopsin, and other hormone and neurotransmitter receptors. Rhodopsin contains a covalently bound chromophore, 11-*cis* retinal, that isomerizes to the all-*trans* configuration upon interaction with light. The change in chromophore structure is propagated through the protein to effect GTP exchange by its partner G protein, transducin. The structure of inactive (dark) bovine rhodopsin was first described 8 years ago,[4] but this is an unusual GPCR due to its relative stability and abundance from natural sources. Other GPCRs are far less abundant, so structural studies require overexpression of recombinant material.

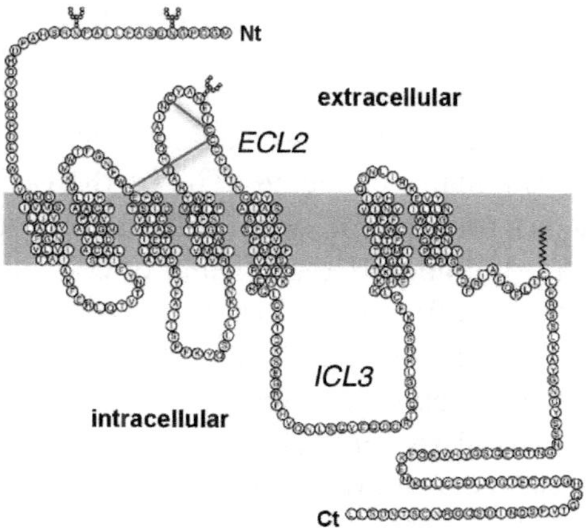

Figure 1. **Primary structure of the β_2AR.** The membrane bilayer is indicated by the blue bar. The third intracellular loop (ICL3) responsible for interaction with G proteins is indicated. Disulfide bonds in the second extracellular loop (ECL2) are shown as red lines. N-linked carbohydrates are indicated by the small circles.

In addition to its natural abundance, rhodopsin is unusual because in the dark it does not catalyze guanine nucleotide exchange in transducin. In contrast, β_2AR and other GPCRs that bind to diffusible ligands display a baseline level of exchange activity in the absence of an activating ligand.[5] This observation implies that the receptor is in a conformational equilibrium amongst states that do or do not activate its associated G protein, a notion supported by a number of biophysical studies.[2] In general, GPCR ligands are classified according to their activity relative to their unliganded, basal state. *Agonists* are compounds that stimulate GTP exchange activity over baseline, such as isoproterenol in β_2AR. *Partial agonists* stimulate exchange to a lower level, even under saturating conditions. *Inverse agonists* suppress exchange activity relative to baseline, whereas *antagonists* neither raise nor lower activity but block binding of other ligands.

Here we describe several approaches that were used to overcome the instability and conformational heterogeneity of β_2AR, and which led to the crystal structure of this molecule. In the first, monoclonal antibodies were prepared to a three-dimensional epitope on β_2AR, and the complex of a Fab fragment bound to the receptor was crystallized in the presence of a strong inverse agonist in lipid bicelles.[5,6] These crystals yielded a structure at an anisotropic resolution of 3.4/3.7 Å. In the second approach, the third intracellular loop was replaced with a small, stable protein, T4 lysozyme (T4L).[7] The β_2AR-T4L chimera was crystallized in the presence of an inverse agonist in a lipidic cubic phase and solved at 2.4 Å resolution.[8]

1.1. FEATURES THAT AFFECT CRYSTALLIZATION

As noted above, β_2AR exists in a conformational equilibrium among activating and non-activating states, so producing a homogeneous population of receptor molecules is a major challenge. The β_2AR was expressed in insect cells using the baculovirus system, with purified, detergent-solubilized protein obtained at a level of about 1 mg per liter of cell culture.[9] An important part of the purification is affinity chromatography on an immobilized ligand, which removes approximately 50% of the molecules and insures that the all of the purified protein is active with respect to ligand binding. It was found that detergent-solubilized receptor is more stable when bound to inverse agonists than to agonists. The inverse agonist carazolol binds to β_2AR with sub-nanomolar affinity and has a very slow off rate ($t_{1/2} > 30$ h), and was found to stabilize the receptor.[7] Note that carazolol is a partial inverse agonist, as there is significant exchange activity displayed by the receptor in the presence of this compound.[5]

Several other features of β_2AR that affect crystallization are highlighted in the primary structure (Fig. 1). First, the presence of heterogeneous and/or flexible carbohydrate moieties can potentially interfere with crystallization. The β_2AR contains three N-linked glycosylation sites. The first two are in the N-terminal peptide before transmembrane helix 1 (TM1), and are well conserved; these could not be removed without compromising expression and were present in the crystallized constructs. There is also a single site in extracellular loop 2 (ECL2) that is not conserved, and this was removed by mutating the Asn to Asp.[9] Second, flexible or disordered regions of proteins oppose the formation of crystal contacts. Removing these regions is frequently essential to produce crystals of water-soluble proteins. Most membrane protein crystals form through contacts between their extramembranous, water-soluble regions, so removing these regions is potentially undesirable. Moreover, for multipass membrane proteins like GPCRs, it is possible to remove flexible regions of the N- and C-termini, but loops connecting helices cannot simply be cut out. In the case of β_2AR, proteolytic sensitivity and fluorescence studies indicate that the last ~40 C-terminal residues are flexible, as is the third intracellular loop (ICL3). ICL3 is required for interaction with G proteins, and spectroscopic studies indicate that the relative arrangement of TM3, TM5 and TM6 changes with activation, indicating that ligand binding and ICL3 are allosterically coupled.[2] Thus, stabilizing a particular conformation of ICL3 and the helices to which it is connected was critical both for expanding the amount of polar surface area for formation of lattice contacts, and also for obtaining a conformationally homogeneous population of receptor molecules for crystallization.

1.2. ANTIBODY STABILIZATION OF ICL3

In order to stabilize the conformation of ICL3 and to increase the water-soluble surface area available to form lattice contacts, monoclonal antibodies were raised against purified β_2AR.[6] Detergent-solubilized β_2AR was reconstituted into phospholipid vesicles at a very high protein:lipid ratio. These vesicles contained randomly oriented β_2AR molecules so that both extracellular and intracellular regions were presented to the immune system. Of the nine monoclonal antibodies obtained, four bound to the extracellular surface of the receptor, and the other five bound to the intracellular side.

An important factor in obtaining an antibody suitable for crystallizing a protein-Fab complex is that the antibody should recognize a three-dimensional epitope, i.e., an epitope formed by the folded structure, rather than a linear peptide, since the latter is likely to be part of a flexible region and will probably not form a spatially homogeneous, well-defined complex. Since the antibodies all recognize the native protein in an ELISA assay, they were tested for binding to denatured β_2AR on a western blot. Two antibodies, 5 and 9, reacted only weakly with the denatured receptor.[6] These results indicate that these antibodies likely bind to a folded, three-dimensional epitope. Next, the antibodies were tested for their effects on the fluorescence of tetramethylrhodamine attached at Cys265, which is near the base of TM6 next to ICL3 and reports on agonist-induced conformational changes.[10] Antibody 5 produced a significant change in fluorescence, suggesting that it was altering the environment of the fluorophore by binding to ICL3. This was confirmed by exploiting the presence of a number of basic resides in ICL3 that give rise to a characteristic tryptic digestion pattern of purified β_2AR. A Fab fragment of antibody 5 was prepared (Fab5) and the receptor was subjected to trypsin in the presence or absence of Fab5. The N-terminal portion of ICL3 was protected by Fab5, which indicated that this antibody binds ICL3.[6]

A fluorescence assay was used to determine the effect of Fab5 on the conformation of β_2AR. By homology to rhodopsin, Ile135 is on the cytoplasmic side of TM3, and Ala271 is near the cytoplasmic end of TM6. The double mutant I135W/A271C can be derivatized at C271 with the fluorophore bimane. Upon activation, TM3 and TM6 move closer together such that the bimane quenches the W135 fluorescence.[11] Using this spectroscopic probe, it was found that Fab5 did not interfere with agonist induced conformational changes, and it had no significant effect on antagonist or inverse agonist binding affinity.[6] Therefore, Fab5 does not appear to alter any properties of the wild-type receptor.

2. Crystal structure of the β_2AR-Fab5 complex

The carazolol-bound β_2AR-Fab5 complex was crystallized in lipid bicelles made from DMPC and the detergent CHAPSO, using ammonium sulfate as the precipitant. The crystals were quite small, with typical dimensions 150 x 20 x 5 μm^3, and the largest samples were 300 x 30 x 10 μm^3. Only limited diffraction could be observed on conventional synchrotron beamlines, but using very bright microfocus beamlines (ESRF ID13 and APS 23ID) the crystals showed diffraction to nearly 3 Å. The space group is C2, a = 338.4 Å, b = 48.5 Å, c = 89.4 Å, β = 104.6° with one β_2AR-Fab5 complex in the asymmetric unit. The crystals were very radiation sensitive, but given their shape complete data could be obtained from a single crystal by measuring a few degrees of data at one position and then translating to a fresh volume of the crystal. An initial data set from full-length β_2AR bound to Fab5 was measured at ESRF to 4.1 Å, and was used for molecular replacement calculations. Crystals were improved by truncating β_2AR at residue 365 (equivalent to the last residue of rhodopsin), and complete data from these crystals were obtained at ESRF. The data are highly anisotropic, extending to 3.4 Å in the plane of the membrane but only 3.7 Å perpendicular to the plane of the membrane.

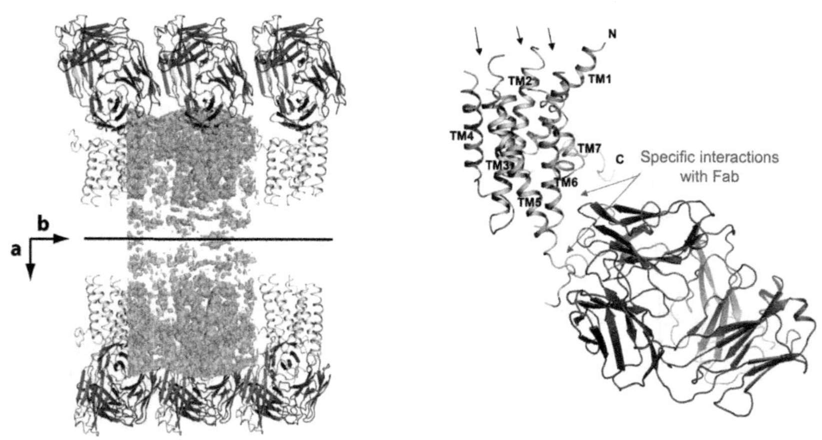

Figure 2. **Structure of the β_2AR-Fab5 complex.** The heavy and light chains of the Fab are shown in blue and red, and the receptor in gold. *Left,* view of crystal packing. The 0.7 σ contour of the 2Fo-Fc map is shown in grey. The crystallographic a and b axes are indicated; the extracellular sides of the receptors pack around the twofold axis indicated by the horizontal line. *Right,* overall structure of the complex. Black arrows point to the missing extracellular loops, and β_2AR residues that interact with Fab5 are highlighted in green, showing that the Fab recognizes a discontinuous, three-dimensional epitope. From [5].

The structure was solved by molecular replacement using Fab constant and variable regions as search models (the former all atom, the latter poly-alanine). Phases made from the rigid-body refined Fab search model produced an electron density map that clearly showed the seven transmembrane helices. Molecular replacement and rigid-body refinement were used to place rhodopsin as a third search model (after the two halves of the Fab), which confirmed our manual interpretation of the density. As expected, the Fab interacts with the cytoplasmic side of β_2AR, and Fab5 molecules mediate many of the contacts in the lattice (Fig. 2). There is also an interaction between the extracellular sides of β_2AR molecules related by a two-fold axis. It appears that the packing in this region is very poor, as the electron density is very weak and the diffraction along a^* is very weak. As a result, the refined structure (R = 0.217, Rfree = 0.269) includes most residues of the Fab, but the entire β_2AR extracellular region is absent, and many side chains are missing in the transmembrane region near the extracellular side of the receptor. We attempted to improve the packing between the extra-cellular regions of adjacent β_2AR molecules by proteolytically removing the first 24 residues, but refinement of the model against data measured from these crystals did not reveal any more of the structure. It is worth noting that despite the lack of experimental phases, the β_2AR model is relatively bias-free, as the well-ordered Fab5 contributes the majority of the scattering and therefore provides independent phase information.

Despite the limitations of the β_2AR-Fab5 structure, several conclusions could be drawn by comparison to rhodopsin. The transmembrane helices superimpose with dark rhodopsin with a root-mean-square deviation (rmsd) of 1.6 Å. Although they are overall similar in structure, the β_2AR helices adopt a more open arrangement near the cytoplasmic side of the membrane. In this region of rhodopsin, TM3 and TM6 are directly linked by a salt bridge between a conserved Arg in TM3 and a Glu in TM6 (Fig. 3). This

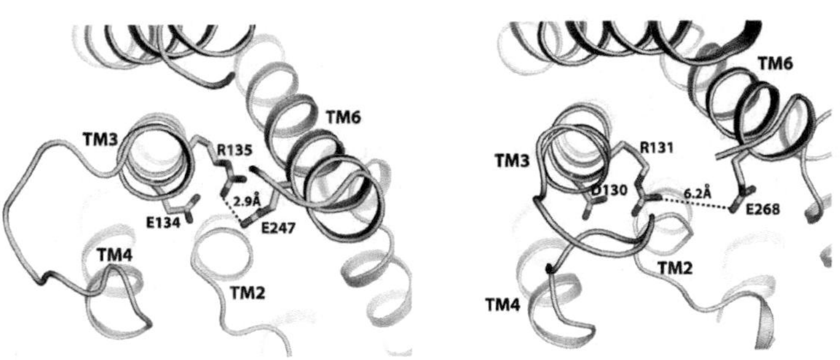

Figure 3. **The ionic lock is broken in β_2AR.** *Left,* view of rhodopsin from the cytoplasmic side showing the ionic lock between R135 and E247. *Right,* equivalent view of β_2AR made by superimposing only TM3 on rhodopsin to highlight the relative movement of TM6. The distance between R131 and E268 is too far to form a salt bridge. From [5].

"ionic lock" is thought to contribute to the inactivity of dark rhodopsin. TM3 and TM6 are farther apart in β_2AR, and the equivalent salt bridge is absent (Fig. 3). This structural difference correlates with the fact that even in the presence of carazolol the β_2AR displays significant activity. In fact, a low-resolution structure of photoactivated rhodopsin also displays opening of this region and breaking of the ionic lock.[12]

After refinement, the only significant residual electron density feature overlaps with retinal in rhodopsin. It was expected from homology and site-directed mutagenesis data of surrounding residues that this is the ligand-binding site, but carazolol could not be modeled with confidence into the electron density. Fortunately, the use of the T4L chimera allowed a complete model of the β_2AR and carazolol to be obtained.

3. Engineering of a β_2AR-T4L chimera for crystallization

The second approach to stabilizing the β_2AR and increasing polar surface area for crystallization was to replace the large, flexible ICL3 with a small, well-ordered protein. T4 lysozyme was chosen as (a) its N- and C-termini are 10.7 Å apart (PDB ID 2LZM), close to the 15.9 Å distance between the ends of TM5 and TM6 in rhodopsin, and (b) it has been crystallized under many conditions and should therefore be amenable to forming lattice contacts.[7] A number of constructs were prepared by varying the length of the junctions between β_2AR and T4L. The construct with the shortest linker that still expressed robustly was chosen in order to minimize the chance that the T4L would be flexibly linked and less likely to crystallize.[7] In addition, the C-terminus was truncated at residue 365 as was done for the β_2AR-Fab5 complex.

Given the importance of ICL3 in the allosteric mechanism that links ligand binding to G protein activation, it was essential to characterize the pharmacology of the β_2AR-T4L chimera to insure that replacing the loop did not fundamentally alter the structure of β_2AR. It was found that the fusion protein has wild-type affinity for antagonists and inverse agonists, but it binds agonists 2–3x more strongly than the wild-type receptor.[7] This property is a hallmark of constitutively active mutants of β_2AR, which display elevated basal activity in the absence of ligands. A fluorescence assay in which exposure of bimane attached to a native Cys residue at position 265, which becomes more solvent exposed upon ligand-induced activation, was used to confirm that the β_2AR-T4L protein undergoes normal conformational changes upon agonist binding. Thus, the replacement of ICL3 with T4L does not appear to fundamentally alter the structure of the remainder of the receptor, making it a good crystallization target.

4. Crystal structure of β_2AR-T4L

The β_2AR-T4L chimera bound to carazolol was crystallized in lipid bicelles as described for the β_2AR-Fab5 complex, but the crystals did not diffract very well. In parallel, the protein was crystallized in a lipidic cubic phase (LCP), exploiting a robotic system that allowed setting up LCP crystallization trials in small volumes. Very small (average size 30 x 15 x 5 μm^3) crystals were obtained in an LCP of monoolein doped with 8–10% cholesterol.[8] Using the microfocus beamline on APS 23ID, diffraction was observed to 2.2 Å. The space group is C2, a = 106.3 Å, b = 169.2 Å, c = 40.2 Å, β = 105.6°, with one β_2AR-T4L in the asymmetric unit. The crystals were very radiation sensitive, and given their size data had to be collected from multiple crystals. A complete 2.4 Å data set was obtained from 27 crystals.[8] The structure was solved by molecular replacement and refined to R and Rfree values of 0.198 and 0.232. The molecule is very well packed in the lattice, such that the temperature factors are very uniform throughout the structure (Fig. 4). Most importantly, the entire sequence could be built without chain breaks; only the N-terminal 28 and C-terminal 23 residues are disordered. In addition to the protein, the structure includes carazolol, a palmitate molecule covalently attached to Cys342, three molecules of cholesterol, and a number of solvent and crystallization additive molecules.[8]

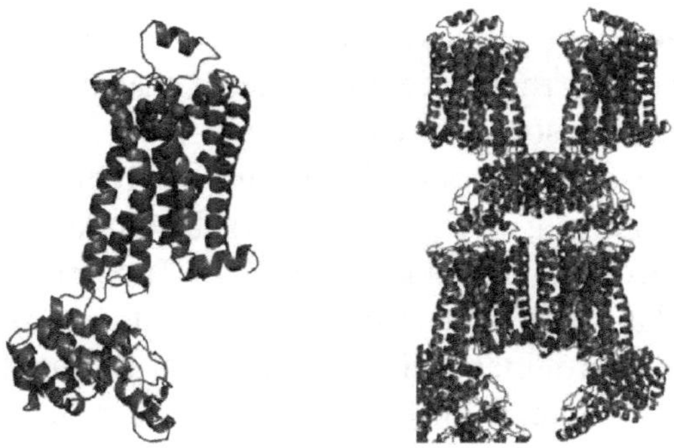

Figure 4. **Structure of β_2AR-T4L.** β_2AR is shown in blue, with carazolol atoms shown as red spheres. T4L is shown in grey. *Left*, overall structure. *Right*, packing in the crystal lattice. From [8].

The T4L moiety mediates most of the contacts between molecules in the lattice[8] (Fig. 4). The one exception is a contact between two receptor molecules around a crystallographic twofold axis, which is largely mediated by ordered lipids, both cholesterol and palmitate. The significance of this contact remains to be investigated.

Comparison of β_2AR-T4L with the wild-type receptor structure yields an rmsd of 0.8 Å for the transmembrane helices visible in the β_2AR-Fab5 structure, which given the low resolution of the latter, indicates that the β_2AR essentially identical in the two structures.[7] This represents a structural validation of the chimera strategy. The only significant difference is at Phe264, which packs between two helices in the wild-type structure but is "flipped out" to interact with lysozyme in the chimera. The ionic lock is also broken the in chimera, although in this case the TM6 Glu268 (Fig. 3) forms a salt bridge with an Arg residue in T4L. Although this could be considered an artifact, the fact that the wild-type receptor also shows a broken ionic lock strongly suggests that this is a genuine feature of carazolol-bound receptor.

4.1. COMPARISON TO RHODOPSIN

The transmembrane helices of rhodopsin superimpose on those of β_2AR-T4L with an rmsd of 2.7 Å.[8] This is larger than that noted above for the wild-type receptor, but it reflects the fact that more of the structure is visible and that there is more divergence on the extracellular side of the TM region than near the intracellular side. The 2.7 Å rmsd is larger than might be expected based strictly on sequence considerations, which may reflect not only sequence differences but also where the two structures sit on the spectrum of activation – dark rhodopsin is completely inactive, whereas carazolol-bound β_2AR retains has some activity.

The most striking difference between β_2AR and rhodopsin is the structure of the extracellular region (Fig. 5). In rhodopsin, the N-terminal peptide and the second extracellular loop (ECL2) combine to form a four-stranded β sheet that sit over the retinal-binding site and shield retinal from the extracellular environment.[4] In β_2AR, the N-terminal peptide is disordered, and ECL2 contains an α helix that is held in place by disulfide bonds and hydrophobic packing interactions.[8] As a result, the ligand-binding site is relatively accessible to the extracellular environment (Fig. 5), which likely enables diffusion of ligands in and out of the site.

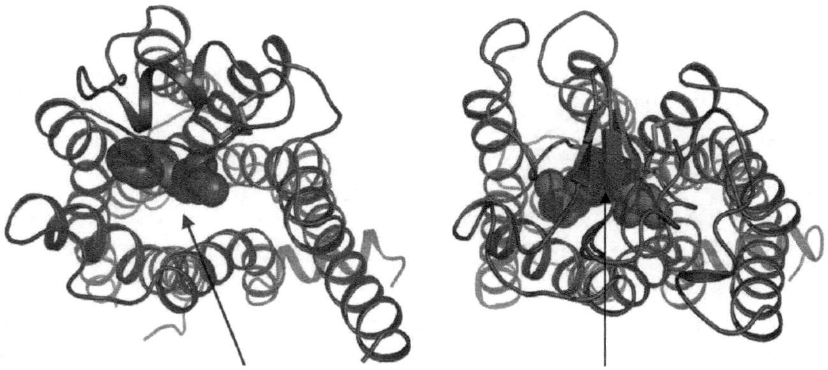

Figure 5. **View of β_2AR (cyan, left) and rhodopsin (magenta, right) from the extracellular side.** The arrows point to the ligand-binding (β_2AR) or retinal-binding (rhodospin) site.

4.2. LIGAND-BINDING SITE AND COUPLING TO ICL3

Carazolol is bound in a pocket equivalent to that of the retinal-binding site in rhodopsin[7,8] (Fig. 6). The charged secondary amine and OH groups common to all β_2AR ligands interact with conserved polar residues present on TM3 and TM7. The carbazole ring forms extensive packing interactions with hydrophobic aliphatic and aromatic residues on TM 3, 5, and 6, and the carbozole NH interacts with Ser203 of TM5. Many of the polar β_2AR residues involved in these interactions were previously identified by mutagenesis studies as important for interaction catecholamine β_2AR ligands. However, several other residues identified in this manner do not interact with the bound carazolol (Fig. 6). Therefore, the agonist isoproterenol was modeled into the site by superimposing the common amine and OH moieties onto those of carazolol, and the catechol ring was then moved by rotation about the one free bond. No contacts with the residues identified to be important for interaction with the catechol ring were observed.[7] This observation suggests that the ligand-binding site is altered in the presence of an activating ligand, consistent with the notion that different pharmacological ligands stabilize different conformations of the receptor.[2,10,11,13,14]

At the base of the carazolol binding site is a conserved Trp residue, W286 (Fig. 6). This position is two residues N-terminal to a conserved proline that kinks TM6, and is though to be important to the movement of helices needed for activation.[15] It is thought that the rotomer of this Trp residue changes in response to ligand binding, and this change propagates to

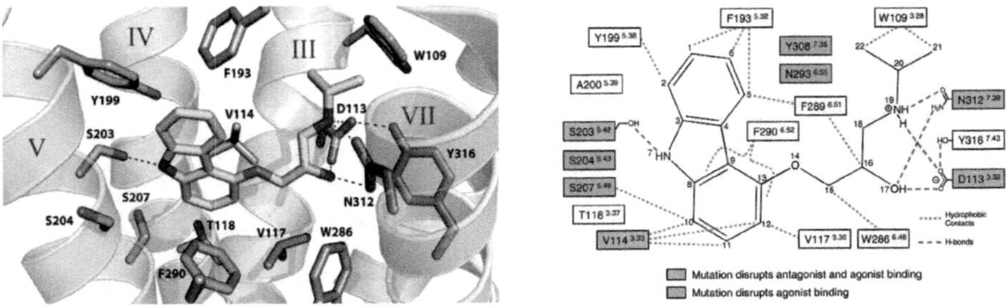

Figure 6. **Carazolol binding to the β_2AR.** *Left*, 3-dimensional diagram of carazolol (yellow) binding to β_2AR (grey). Nitrogen and oxygen atoms are shown in blue and red, respectively. *Right*, schematic diagram of the interactions. Mutation of residues highlighted in dark green boxes selectively disrupt agonist binding, whereas mutation of residues highlighted in light purple affects binding of both antagonist and agonists. From [7].

ICL3 and the bound G-protein. To gain insight into the nature of conformational coupling between the ligand-binding site and ICL3, the G-protein interaction site, the positions of constitutively active mutants (CAMs) and uncoupling mutants were mapped onto the structure. CAMs display elevated levels of basal activity in the absence of ligands, which likely means that these positions stabilize the inactive state of the receptor. For example, the mutation L272C produces a CAM phenotype. Leu272 forms a number of packing interactions with residues on TM3 and TM5,[5] so introduction of the smaller Cys residue probably loosens the packing and lowers the transition barrier to the activated state. Uncoupling mutations (UCMs) uncouple agonist binding from G protein activation. These residues are not required for ligand binding but are required for the stability and/or function of the active state.

Both CAM and UCM positions are outside of the ligand-binding site[7] (Fig. 7), so these residues participate in the allosteric coupling between the ligand-binding and G-protein interaction sites. CAMs are centrally located on TM3 and TM6 (Fig. 7), helices that spectroscopy indicates move in response to activating ligands.[11] The UCM positions are more widely distributed, but most are near the cytoplasmic side of the TM region, with a cluster of UCM positions present at the cytoplasmic end of TM7 (Fig. 7). Although these residues do not contact directly CAM positions, the two sets of residues are linked by packing interactions (Fig. 7). Thus, it is clear that the movements of one set are linked to the other, so in a general sense it is clear that changes in the ligand-binding site are propaged through these residues to ICL3.

In addition to packing interactions, there is a water-filled cavity in the cytoplasmic half of the receptor formed by conserved polar residues from TM2, 3, 6 and 7[7] (Fig. 7). Remarkably, both the polar side chains and ordered water molecules overlap closely with those found in dark rhodopsin. Thus, this cavity appears to be a conserved structural feature. A water-filled, loosely packed cavity would facilitate conformational transitions since there are fewer energetic barriers to rearrangement relative to the repacking of non-polar side chains represented by the CAM and UCM positions.

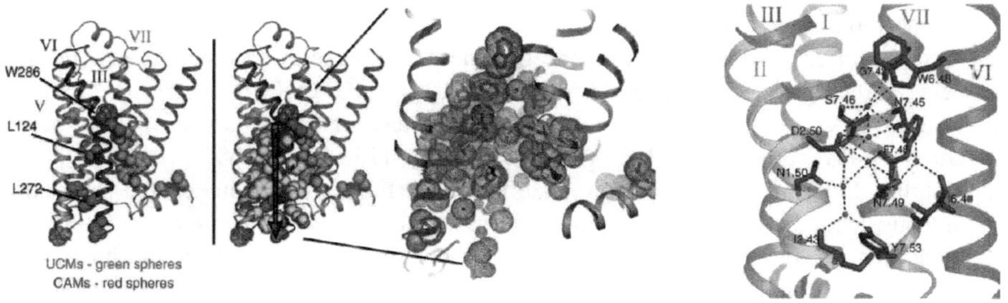

Figure 7. **Coupling of ligand binding and G-protein interaction sites.** *Left,* location of constitutively active mutations (CAMs) (red) and uncoupling mutations (UCMs) (green). Residues that are within 4 Å of two of the CAMs are shown in yellow, highlighting the linked packing of the CAM and UCM sites. *Right*, network of ordered water molecules and polar side chains within the receptor. From [7].

5. Prospects for modeling GPCRs

Ultimately, the structures of more pharmacological states, in particular agonist-bound structures, will be essential for understanding the mechanism that couples ligand binding and G-protein activation. It is hoped that crystal structures along the activation pathway, combined with spectroscopic experiments and molecular dynamics calculations, will enable detailed understanding of receptor activation. Finally, the agonist-bound ligand-binding site must differ significantly from that observed in the carazolol-bound structure. This emphasizes that homology models have to consider both sequence differences as well as the pharmacological state of interest. Until structures of an activated receptor are known, constructing models of the agonist-bound sites of other receptors remains a serious challenge.

References

1. Lefkowitz, R. J. & Shenoy, S. K. Transduciton of receptor signals by beta-arrestins. Science 308, 512–517 (2005).
2. Kobilka, B. K. & Deupi, X. Conformational complexity of G-protein-coupled receptors. Trends Pharmacol Sci 28, 397–406 (2007).

3. Pierce, K. L., Premont, R. T. & Lefkowitz, R. J. Seven-transmembrane receptors. Nat. Rev. Mol. Cell Biol 3, 639–650 (2002).
4. Palczewski, K. et al. Crystal structure of rhodopsin: A G protein-coupled receptor. Science 289, 739–745 (2000).
5. Rasmussen, S. G. et al. Crystal structure of the human beta2 adrenergic G-protein-coupled receptor. Nature 450, 383–387 (2007).
6. Day, P. W. et al. A monoclonal antibody for G protein-coupled receptor crystallography. Nat Methods 4, 927–929 (2007).
7. Rosenbaum, D. M. et al. GPCR engineering yields high-resolution structural insights into beta2-adrenergic receptor function. Science 318, 1266–1273 (2007).
8. Cherezov, V. et al. High-resolution crystal structure of an engineered human beta2-adrenergic G protein-coupled receptor. Science 318, 1258–1265 (2007).
9. Kobilka, B. K. Amino and carboxyl terminal modifications to facilitate the production and purification of a G protein-coupled receptor. Anal Biochem 231, 269–71 (1995).
10. Swaminath, G. et al. Probing the beta2 adrenoceptor binding site with catechol reveals differences in binding and activation by agonists and partial agonists. J Biol Chem 280, 22165–22171 (2005).
11. Yao, X. et al. Coupling ligand structure to specific conformational switches in the beta2-adrenoceptor. Nat Chem Biol 2, 417–422 (2006).
12. Salom, D. et al. Crystal structure of a photoactivated deprotonated intermediate of rhodopsin. Proc Natl Acad Sci U S A 103, 16123–16128 (2006).
13. Ghanouni, P., Steenhuis, J. J., Farrens, D. L. & Kobilka, B. K. Agonist-induced conformational changes in the G-protein-coupling domain of the beta 2 adrenergic receptor. Proc Natl Acad Sci U S A 98, 5997–6002 (2001).
14. Gether, U. et al. Agonists induce conformational changes in transmembrane domains III and VI of the beta2 adrenoceptor. Embo J 16, 6737–6747 (1997).
15. Shi, L. et al. beta2-adrenergic receptor activation. Modulation of the proline kink in transmembrane 6 by a rotamer toggle switch. J Biol Chem 277, 40989–40996 (2002).

CAN STRUCTURES LEAD TO BETTER DRUGS? LESSONS FROM RIBOSOME RESEARCH

ADA YONATH*
Department of Structural Biology, Weizmann Institute, Rehovot, Israel

Abstract. Ribosome research has undergone astonishing progress in recent years. Crystal structures have shed light on the functional properties of the translation machinery and revealed how the ribosome's striking architecture is ingeniously designed as the framework for its unique capabilities: precise decoding, substrate mediated peptide-bond formation and efficient polymerase activity. New findings include the two concerted elements of tRNA translocation: sideways shift and a ribosomal-navigated rotatory motion; the dynamics of the nascent chain exit tunnel and the shelter formed by the ribosome-bound trigger-factor, which acts as a chaperone to prevent nascent chain aggregation and misfolding.

These linkage between these findings and crystal structures of ribosomes with over two dozen antibiotics targeting the ribosome, most of which of a high therapeutical relevance, illuminated various modes of binding and action of these antibiotics; deciphered mechanisms leading to resistance; identified the principles allowing for the discrimination between pathogens and eukaryotes despite the high ribosome conservation; enlightened the basis for antibiotics synergism, namely the conversion of two weakly acting compounds to a powerful antibiotic agent; indicated correlations between antibiotics susceptibility and fitness cost and revealed an novel induced-fit mechanism exploiting ribosomal inherent flexibility for reshape the antibiotic binding pocket by remote interactions.

Keywords: Crystal structure, catalytic mechanism, antibiotics, protein synthesis, translation machinery, RNA, ribonucleoprotein, enthalpy driven binding, entropy

* To whom correspondence should be addressed. Ada Yonath, Department of Structural Biology, Weizmann Institute., Rehovot, Israel; e-mail: ada.yonath@weizmann.ac.il

J.L. Sussman and P. Spadon (eds.), *From Molecules to Medicines*,

1. Introduction

An adult human body contains approximately 10^{14} cells, each containing about a billion proteins. Proteins are constantly being degraded, and simultaneous production of proteins is therefore required. The translation of the genetic code into proteins is performed by a complex apparatus comprising the ribosome, messenger RNA (mRNA), transfer RNAs (tRNAs) and accessory protein factors. The ribosome, a universal dynamic cellular ribonucleoprotein complex, is the key player in this process, and typical mammalian cells can contain over a million ribosomes (the 'factories' that translate the genetic code into proteins). Even bacterial cells contain ~100,000 ribosomes. Many ribosomes act simultaneously along the mRNA, forming superstructures called polysomes. They act as polymerases synthesizing proteins by one-at-a-time addition of amino acids to a growing peptide chain, while translocating along the mRNA template. In bacteria, ribosomes produce proteins on a continuous basis at an incredible speed of >15 peptide bonds per second.

Ribosomes are composed of two subunits (Table 1); comprising long chains of ribosomal RNA (rRNA) in which many ribosomal proteins (r-proteins) are entangled. The ratio of 2:1 for rRNA:r-proteins is maintained throughout evolution, with the exception of the mammalian mitochondrial ribosome in which almost half of the bacterial rRNA is replaced by r-proteins. Despite the size difference (Table 1), ribosomes from all kingdoms

TABLE 1. Biophysical and chemical characterization of ribosomes.

	Prokaryotic ribosome	Eukaryotic ribosomes
Sedimentation coefficient	70S (~2.4 MDa)	80S (~4 MDa)
Small subunit	30S - One rRNA molecule (16S with ~1500 nucleotides) ~ 21 different proteins (S1–S21)	40S - One rRNA molecule (18S with 1,900 nucleotides) ~ 33 different proteins (S1–S33)
Large subunit	50S - Two rRNA molecules (5S and 23S, with ~120 and ~2900 nucleotides, respectively) ~ 31 different proteins (L1–L31), among which only L12 is present in more than a single copy	60S - Three rRNA molecules (5S, 5.8S and 28S, with 120, 156 and 4,700 nucleotides, respectively) ~ 50 different proteins (L1–L50)

of life are functionally conserved; with the highest level of sequence conservation in the functional domains. Comparisons of rRNA sequences of widely diverged species and extrapolation of structures from eubacteria via archaea to eukaryotes indicate that the largest structural differences are at the periphery, away from the central core.

2. Recent progress in ribosomal crystallography

Remarkable accomplishments in characterizing the machinery of protein biosynthesis have been made at the turn of the millennium. Following two decades of preparative efforts [1], structures of ribosomal particles have been determined. These include the large ribosomal subunit of the archaeon *Haloarcula marismortui*, H50S [2] and the eubacterium *Deinococcus radiodurans*, D50S [5], the small subunit from the eubacteria *Thermus thermophilus*, T30S [4, 5] and the entire ribosome from the same source, T70S [6]. The earlier studies are reviewed extensively (e.g. [7–9]). More recent structures include vacant ribosome [10], functional complexes of ribosomes with mRNA and tRNAs [11–15] and/or with recycling [16, 17] and release factors [18]. Additional crystal structures are of functional complexes of small subunits with mRNA [19] and modified tRNAs [20, 21]; large subunits with substrate analogs extending from the initial (e.g. [22]) to more sophisticated complexes [23, 24]; large subunit with non-ribosomal auxiliary factors: the first chaperone to encounter the emerging nascent protein, the trigger factor [25–27] and the ribosomal recycling factor [28]. Most of the currently available structures are of ribosomes from organisms that have adapted to extreme environments, as these are more suitable for crystallization. Yet, owing to the high level of conservation of the ribosomal functionally relevant domains, the extremophile ribosomes and their genetically modified phenotypes can represent ribosomes from non-extremophile species [29].

Stimulated by the emerging structures, ribosome research has undergone a quantum jump, yielding exciting findings concerning various aspects of protein biosynthesis in prokaryotes (e.g. [30–53]), which could be extended and/or paralleled with corresponding events in eukaryotes (e.g. [54, 55]). Likewise, the structural basis for clinical relevance of antibiotics targeting ribosomes despite their high conservation has progressed significantly. Crystal structures of complexes of ribosomal particles with their antibiotics obtained until 2005 have been reviewed elsewhere (e.g. [56–60]. More recent findings are reported in [61, 67] or presented here. Still emerging are elaborate analyses of results that have led to plausible [68] or controversial biological implications. An example for the latter is the finding that mutation of the nucleotide determining macrolide antibiotic binding to eubacterial ribosomes (2058) form guanine, as in eukaryote, to adenine, as in pathogen [70] results in antibiotic binding, but does not confer antibiotics sensitivity [71] as originally expected [70].

Account of the currently available crystallographic data and highlights of some of the issues that remain unresolved, alongside a brief summary of the functional implications of the recent structures of the bacterial ribosomes are presented in this review. The bacterial ribosomes are of immense contributions to the understanding the universality of protein biosynthesis and the divergence from it. Thus, although the translation apparatus in eukaryotes is larger and more complicated than in bacteria, the research on the bacterial ribosome has led to imperative insights into key issues concerning ribosomes of the eukaryotic kingdom as well as opened new routes for the development and improvement of ribosomal antibiotics. These are accompanied by several (out of many) of the recently published numerous biochemical, genetic and cryo-EM studies that expand ribosome research beyond the crystal structures.

3. Ribosome mode of action

Ribosomes comprise two ribonucleoprotein subunits (Figure 1a) that associate to form the functional ribosome. While elongation proceeds, each subunit operates cooperatively. The small subunit provides the mRNA binding machinery (Figure 1b) and the path along which the mRNA progresses, the decoding center and the major component controlling translation fidelity. The large subunit performs the main ribosomal catalytic function, namely amino acid polymerization, and provides the protein exit tunnel. tRNAs, the molecules decoding the genetic information and carrying the amino acids to be incorporated in the growing protein, are the non-ribosomal entities that join the two subunits, as each of their three binding sites: A-(aminoacyl), P-(peptidyl), and E-(exit) reside on both subunits (Figure 1a). The initial tRNA binds to the first codon of the mRNA at the P-site and the next tRNA, which enters the ribosome via the dynamic L7/12 stalk (Figure 1a), attaches to the next codon at the A-site. While a peptide bond is formed, the A-site tRNA is translocated to the P-site and the deacylated tRNA moves from the P-site to the exit (E)-site on its way out from the ribosome, through the mobile L1 stalk (Figure 1a). At each elongation cycle both subunits participate in translocating the mRNA and the tRNA molecules by a single codon.

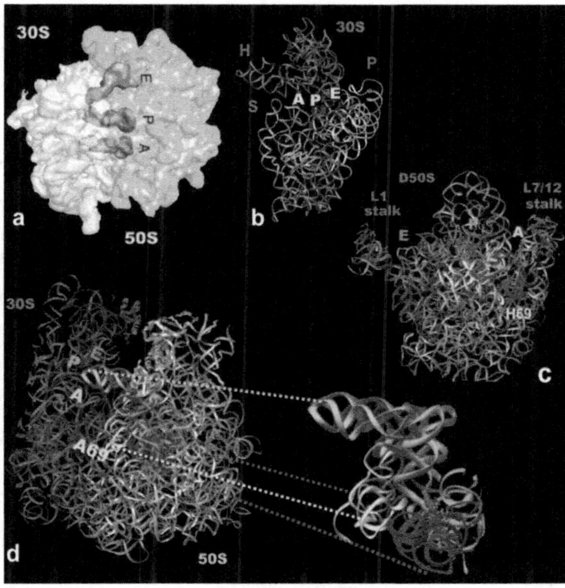

Figure 1. The ribosome functional centers. (a) The two ribosomal subunits. Left: the small ribosomal subunit (T30S) [4]. The approximate positions of codon-anticodon interactions of A-, P- and E- tRNAs are shown and the main functional domains are indicated. H, head; S, shoulder; P, platform; L, latch. The arrows designate the approximate directions of the coordinated motions associated with mRNA binding and translocation. The left arrow indicates the creation of the mRNA pore, i.e. the latch motion [4]. Right: The large ribosomal subunit (D50S) [2]. Regions that are involved in amino acid polymerization are indicated. These include the two stalks controlling the A-site tRNA entrance (L7/L12) and the E-site tRNA exit (L1), which are known to undergo a coordinated lateral movement during elongation; the positions where the acceptor stems of the three tRNA molecules (A-, P- and E-) interact with this subunit. Insert: a tRNA molecule on which its two functional domains (the anticodon loop and CCA 3'end, which binds the incoming amino acid or the newly born protein) are marked. The brown circle indicates the portion of the tRNA molecule interacting with the small subunit, and the blue circle shows the portion bound to the large subunit. (b) The positions of initiation factor 3 (IF3) and Shine Dalgarno (SD) region on the small subunit. The small ribosomal subunit is shown in grey. The arrow indicates the possible motion of IF3 C-terminal domain (IF3C). Top: a space-filled view similar to that shown in Figure 1a. Bottom: a more detailed representation of the opposite view. Marked are the IF3 domains (C-terminal, N-terminal and the linker between them); the SD region; the anticodon loops of the three tRNAs (A, P, E), and the proteins involved in IF3 binding. (c) The central location of the symmetrical region in the large ribosomal subunit from D50S, shown in grey, with A- and P-site tRNAs (docked according to [6]) and the symmetrical region (colored blue and green) with its extensions (shown in gold). The symmetrical region is shown by blue and green (for A- and P- sites, respectively) with the pseudo twofold imaginary axis in red. Note that it connects directly or through its extensions (shown in gold) all the large subunit functional regions, including the bridge, connecting it to the decoding site on the small subunit [39, 40].

The surface of the intersubunit interface is composed predominantly of ribosomal RNA (rRNA), and in the assembled ribosome all functional sites are located close to this interface. Hence, unlike typical polymerases, which are protein enzymes, RNA is the major player in ribosome activities. The site of peptide bond formation, the peptidyl transferase center (PTC), is positioned within a universal pseudo twofold symmetrical region (Figure 1c), composed of highly conserved nucleotides and called 'the symmetrical region'. This means that each point on the fold 90 nucleotides comprising one half of the symmetrical region, is related by a rotation of 180° around an imaginary axis, located at the middle of the PTC, to its mate on the other half, which is also composed of 90 nucleotides. In addition to the rRNA fold, this internal symmetry relates the nucleotide orientations (Figure 1d and 2a–c), but not nucleotides sequences. The entire symmetrical region is highly conserved [39, 40] in which 98% of the nucleotides are 'frequent' (found in >95% of sequences from 930 different species from the three domains of life), whereas only 36% of all *E. coli* nucleotides, excluding the symmetrical region, can be categorized as such. Importantly, 75% of the 27 nucleotides lying within 10 Å distance from the symmetry axis are highly conserved. Among them seven are completely conserved [40].

The high level of conservation of the symmetrical region, its central location and its link to all ribosomal features involved in amino acid polymerization (Figure 1c) [7, 23, 39, 40] indicates that it can serve as the element signaling between remote ribosomal locations (up to 200 Å away from each other) and thus can coordinate translation processes. This is consistent with the observed relationship between PTC occupation and mRNA binding to the small subunit [48].

The ribosome is a dynamic molecular machine that involves structural rearrangements as an integral part of the translation machinery. Various motions have been detected by investigating the reasons for disorder in functionally relevant regions in crystals grown under far from physiological conditions [2, 22] or by cryo electron microscopy (e.g. [72]) and single particles methods [48]. In addition, interpolation between the structure of the unbound large subunit, D50S (e.g. [3]) and that of the entire ribosome, T70S, with three tRNAs [6] identified fundamental motions, like the coordinated movement of the two large subunit stalks (Figure 1a) [3, 49, 34] involved in the entrance and release of the A- and E-tRNAs. Also detected in the 30S structure are the head-shoulder movement upon A-site occupation [8] and the 30S head-platform correlated motions (Figure 1a) enabling guidance to mRNA progression [4, 10, 13] together with elongation factor EF-G [73] as part of the ratchet-like intersubunit reorganization [74]. Additional motions were correlated with tunnel gating [68], possible trafficking of nascent chain progression [25], rearrangements caused by elongation factor EF-Tu ternary complex binding that are linked to fidelity control [8], motions within the PTC correlated with activation/deactivation [30], inhibitory action of antibiotics [63] and the rotatory component of the substrate's translocation [7, 23, 39].

4. On the functional contribution of ribosomal proteins

Over the years, the views on the contribution of the ribosomal proteins (r-proteins) to ribosome function have changed dramatically. Originally, r-proteins were thought to carry out the ribosomal catalytic tasks [75], but later it was shown that rRNA performs most of the ribosome functions. The high resolution crystal structures show that in addition to their peripheral globular domains, almost all r-proteins possess elongated loops or terminus extensions, penetrating into the rRNA core, thus seem to serve as entities stabilizing the rRNA conformation. However, alongside their stabilization roles, some r-proteins can facilitate functions requiring mobility (reviewed in [7]). For example, protein L22 appears to cause transient tunnel blockage [68] and L1 and L12, the main protein component of the dynamic L1 and the L7/L12 stalks of the large subunit (Figure 1a) seem to be involved in tRNA translocation (reviewed in [7–9]). Additionally, proteins situated in proximity to functional regions were proposed to support specific activities. Thus, proteins S5, S6 and S12 assist mRNA binding fidelity [8], and proteins L27 [11, 35] (which does not exist in the archaeon H50S) and L2 [76] were suggested to affect peptidyl transferase activity. S12 and L2 are among the few proteins that reside partially on the intersubunit interface and can support the biosynthetic process. Importantly, computational methods found that S12 and L2 are among the most ancient ribosomal proteins [77].

5. Non ribosomal compounds involved in initiation and elongation

tRNA molecules decode the genetic information by matching the complementary bases of their anticodon loop with the codon on the mRNA. All tRNAs are double helical L-shape molecules, except for their anticodon loop and the single stranded 3'end (almost universally CCA) to which the cognate amino acid or the growing peptidyl chain is bound (Figure 1a).

Three non-ribosomal protein factors are involved in the initiation. Initiation factor 2 (IF2) is a GTPase that binds preferentially to initiator tRNA. It acts in a cooperative manner with initiation factor 1 (IF1), which occludes the ribosomal A-site at the small subunit (Figure 1b) and flips out two functionally important bases (A1492 and A1493). These localized changes lead to global alterations in the 30S conformation [8], which seem to be essential for the next steps in translation. Initiation factor 3 (IF3) interferes with subunit association and promotes the ribosome fidelity at the initial phase, by assisting the selection of the initial P-site codon–anticodon interactions. The crystal structure of the C-terminal domain (IF3-c) in complex with T30S indicates its binding to a region proximal to the mRNA channel [8], in a mode suggestive of exploiting its inherent flexibility for an over-the-platform swing to a location suitable for facilitating subunit dissociation (Figure 1b) [56]. Interestingly, IF1 and IF2 (a/eIF1A and a/eIF5B in eukaryotes) are conserved across all three kingdoms of life and cryo-EM studies suggest that they interact with the 30S in a similar manner, although initiation in eukaryotes and archaea requires additional factors.

In prokaryotes, the elongation cycle is driven by GTPase activity of elongation factors. Tu (EF-Tu) delivers the cognate aminoacylated-tRNA to the ribosomal A-site as a ternary complex with GTP, induces long- and short-range conformational alterations, and dissociates after GTP hydrolysis. EF-G contributes to bias the translocation in the forward direction [73]. It binds preferably to the ribosome at its ratcheted conformation, obtained by a rotation of the small subunit relative to the large subunit in the direction of the mRNA movement [74], thus facilitating GTP hydrolysis. Both EF-Tu and EF-G bind to the mobile L7/L12 entrance stalk (Figure 1a) via a conserved region of protein L12 C-terminal domain [49]. In concert with these motions, the deacylated tRNA at the E-site moves towards protein L1, on the other side of the ribosome (Figure 1a), and consequently this protein undergoes a significant conformational alteration in order to release it [2, 6, 10–12].

6. Initiation, subunit association, decoding, and translocation

A prerequisite for correct translation is accurate positioning of mRNA on the ribosome. This step is of utmost importance, hence any divergence can destabilize tRNA binding and inhibit canonical translation initiation [61, 62]. In prokaryotes, mRNA placement is assisted by a target pyrimidine-rich region ('anti Shine-Dalgarno'), located at the 3'end of the 16S RNA. This region anchors the complementary purine-rich sequence at the 5'-end of mRNA ('Shine-Dalgarno' or SD) by numerous interactions (Figure 1b) [13] and creates a chamber for transient stabilization of this otherwise labile double helix [19]. In eukaryotes, mRNA placement requires highly sophisticated machinery [54, 55], and throughout evolution it has involved various non-ribosomal factors.

Crystal structures of prokaryotic ribosomes imply that mRNA entrance to its groove on the small subunit involves a latch-like closing/opening mechanism [4, 6, 15]. These structures also suggest that the mRNA kinks between the A- and P-sites at the decoding region [4, 6], and that this conformation seems to be stabilized by a metal ion, which delineates the border between the two sites and prevents uncontrolled mRNA sliding [11]. Once mRNA and initiator P-site tRNA bind to the small subunit the two subunits associate to form the functional ribosome. The surface complementarily is stabilized by over a dozen intersubunit bridges formed by conformational changes of the interface components [3, 6, 11, 12]. Several bridges seem to play roles beyond merely guaranteeing correct subunit interactions. Among them, bridge B2a is particularly important as it connects the immediate environments of the PTC with the decoding center and has the ability to adopt several conformations, depending on the ribosome functional state [2, 23].

The elongation cycle is composed of decoding, peptide bond formation, amino acid polymerization, detachment of the P-site tRNA from the growing polypeptide chain and release of the deacylated tRNA. These processes are facilitated by translocation, which is a successive coordinated movement of the mRNA and its associated tRNAs through the ribosome from A-site to the P-site and then to the E-site, by one codon at a time (in 3' to 5' direction). Decoding fidelity, namely avoiding disparity between the mRNA codons and the tRNA anticodons is vital for guaranteeing translation accuracy. The incoming aminoacylated-tRNAs are selected for forming the codon- anticodon base pairing with an error rate of 10^{-3}–10^{-4} at the highly conserved RNA-rich decoding center of the small ribosomal subunit. The ribosome plays a major role in this selection, exploiting the inherent flexibility of the decoding center for strictly monitoring the base pairing at the first two positions of each codon, but tolerating non-canonical base pairs at the third position [8]. Furthermore, it appears that normal triplet pairing is not an absolute constraint of the decoding center. For example, flexible expanded anticodon loops of frameshift promoting tRNAs can adopt conformations that allow three bases of the anticodon to span four mRNA bases [20].

The current integrated model for decoding proposes that tRNA selection hinges on discrimination based on the interactions between the ribosomal rRNA and the minor groove of the codon–anticodon duplexes, with a potential to lead to domain closure. Cognate tRNA binding induces global structural rearrangements by domain movements and these modify the conformation of the universally conserved decoding regions so that bases residing in it can interact with the first two base-pairs of the codon-anticodon helix.

7. Peptide bond formation and the polymerase activity of the ribosome

All ribosome crystals structures indicate that the major player in ribosomes activities is RNA [3, 22, 23, 57]. During the past three decades, the preferred substrate analogs used for determining ribosomal functional activity, were 'minimal substrates', namely puromycin derivatives capable of creating a single peptide bond. Using similar compounds, which were believed to act as substrate and transition state analogs for complexes with H50S, it was proposed that four universally conserved rRNA nucleotides catalyze peptide bond formation by a general acid/base mechanism [22]. This proposition was soon challenged by various biochemical and mutational studies (e.g. [30, 31, 43]) and additional crystallographic studies on complexes of H50S with similar, albeit more sophisticated, substrates analogs (e.g. [24]) illuminated several aspects of peptide bond formation, such as conformational rearrangements that the PTC can undergo, but did not lead to a feasible consensus mechanism This could be linked to the finding that in all structures of H50S and its complexes with substrate analogs, almost all regions involved in ribosome function are disordered (namely posses simultaneously multiple conformations) presumably owing hey were constructed under far from physiological conditions [2, 22]. Consequently although these structures did not yield the mechanism of peptide bond formation they illuminated an important aspect in cellular regulation of ribosome function, namely that disorder of functionally relevant ribosome regions might represent a common strategy for avoiding non-productive protein biosynthesis.

Structures of a complex of D50S with either an A-site tRNA acceptor stem mimic (composed of 35 nucleotides, including an aminoacylated 3'end, called ASM) [23] obtained under conditions close to those optimized for protein biosynthesis revealed that the acceptor stem of A-site tRNA interacts extensively with the cavity leading to the PTC, and the bond between it and the tRNA 3'end overlaps the symmetry axis (Figure 2d). The high conservation of the components of the symmetrical region, the linkage between the elaborate PTC architecture and the position of the A-site tRNA observed crystallographically [23] indicates that the translocation of the tRNA 3' end is performed by a combination of two independent, albeit synchronized motions: a sideways shift, performed by the overall mRNA/tRNA translocation, and a rotatory motion of the A-tRNA 3'end along a path confined by the PTC walls (Figure 2e). Navigated and guided by the ribosomal architecture, this rotatory motion provides all of the structural elements for ribosome function as an amino acid polymerase, including the formation of two symmetrical universal base pairs between the tRNAs and the PTC [23, 39, 40], a prerequisite for substrate mediated acceleration, rather than acid-base catalysis [32, 33, 43, 51], and for directing the nascent protein into the exit tunnel.

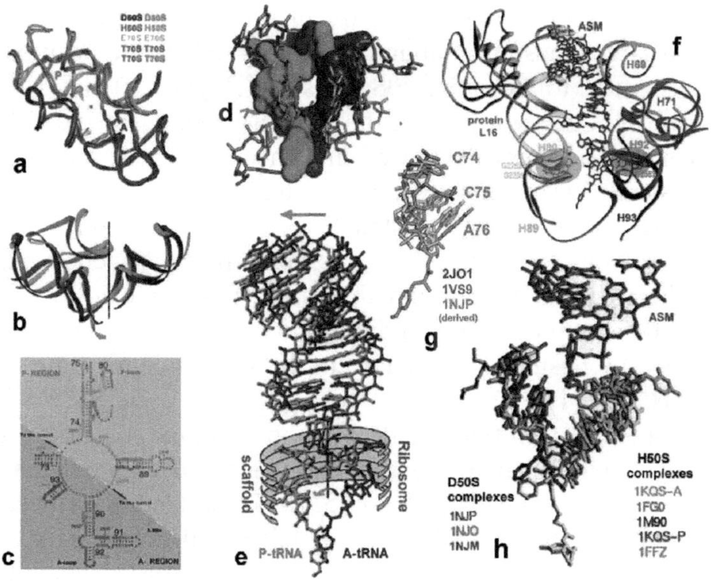

Figure 2. The symmetrical region and peptide bond formation. (a–c). The universal symmetrical region backbone fold. In all structures, the A- and P- sub-regions are shown in blue and green, respectively. The imaginary symmetry axis is shown in red. (a) Superposition of fold of the 180 nucleotides comprising the symmetry region in all known structures, shown as ribbons. The two pseudo-symmetrical sub-regions, containing the A- and the P-sites, are shown in blue and green respectively. The imaginary axis relating the two halves of the symmetrical region is shown as a red rod (or its cross-section). The center of the PTC lies roughly on this axis. (b) Superposition of the backbones of the rRNA comprising the A- and P- sub-regions of the symmetrical region, as obtained by a 180° rotation around the imaginary symmetrical axis, indicating the level of the ribosomal internal symmetry. (c) Two-dimensional representation of the 23S rRNA segment that belongs to the symmetrical region. Symmetrical features are shown in identical colors. (d) Superposition of the locations of short substrate analogs used in crystallographic studies together with H50S and D50S. The PDB accession codes are indicated. (e) The tRNA translocation motion, comprising a synchronized sideways shift, performed as part of the overall mRNA/tRNA sideways translocation (in the direction of the horizontal arrow), and the rotatory motion of the A-tRNA 3'end along a path confined by the PTC grey walls (shown here as ribs). The A-site tRNA and the derived 3'end of the P-site tRNA are shown in blue and green (respectively). The direction of the rotatory motions is indicated by a blue-green curved arrow, the imaginary twofold symmetry axis is red, and the approximated positions of the symmetrical basepairs [23, 32, 39, 40] are shown in yellow. (f) Superposition of the derived P-site CCA (from ASM 3'end by the rotatory motion) on the crystallographically determined locations of the P-site CCA in crystals of 70S complexes [11, 12]. The PDB accession codes are indicated.

Remarkably, the position of the 3' end of P-site, derived by the rotatory motion that was suggested based on the mode of binding of a tRNA mimic to unbound large ribosomal subunit (D50S), overlaps the positions of full-size tRNAs bound to the entire 70S ribosome (Figure 2f) [11, 12]. Furthermore, all nucleotides involved in this rotatory motion of the tRNA 3'end have been shown to be essential by a comprehensive genetic selection analysis [45]. Consistently, quantum mechanical calculations, based on D50S structural data, indicated that the transition state (TS) for this reaction is being formed during the rotatory motion, and is stabilized by hydrogen bonds formed between the rotating moiety and the same rRNA nucleotides [46]. The location of the computed TS is similar to that observed crystallographically for a chemically designed TS analog in the large subunit from a different ribosome, H50S [24].

Differences between full-size tRNAs and 'minimal substrates' were also obtained by biochemical mutagenesis, kinetics and computational studies (e.g. [30–33, 36, 37, 42–44, 50, 51, 53]). These studies showed that the mechanism of peptide bond formation by full-size tRNAs involves substrate mediated catalysis [32], and require the stereochemistry obtained by the rotatory motion [39]. They also highlighted the importance of accurate positioning of the tRNAs, which can be achieved by full-size tRNA or its mimics containing the acceptor stem nucleotides that interact with proximal ribosomal nucleotides [7, 23]. It is important to note, however, that a symmetrical relationship between the reactants of peptide bond formation has been observed in all known structures of ribosomal complexes (Figure 2d), including 'minimal substrates' requiring additional rearrangements. In principle, suitable systems for studying this machinery should include a full-length A-site tRNA bound to the ribosome. However, although 70S ribosomes complexed with full-length aminoacylated-tRNA were crystallized, A-site tRNA 3' ends could not be detected in any of the electron density maps [11–13]. Hence, the only relevant crystallographic information currently available originates from the structure of the complex of D50S with ASM [23].

The correlation between the rotatory motion and amino acid polymerization n rationalizes the apparent contradiction associated with the location of the growing protein chain, since the traditional biochemical methods for the detection of ribosome activity as well as most of the crystallographic studies were based on minimal substrate analogs designed for producing a single peptide bond. These analogs do not undergo A- to P-site translocation, whereas nascent protein elongation requires this motion. Furthermore, the difference between the formation of single peptide bond by minimal substrates and amino acid polymerization highlights the PTC ability to rearrange itself upon substrate binding [7, 30, 58].

The conservation of the symmetrical region is consistent with its vital functions in intra-ribosomal signaling, peptide bond formation and amino acid polymerization. The preservation of the three-dimensional structure of the two halves of the ribosomal frame regardless of the sequence demonstrates the rigorous requirements of accurate substrate positioning in stereochemistry supporting peptide bond formation. This, as well as the universality of the symmetrical region led to the assumption that the ancient ribosome was made of a pocket confined by RNA chains and that the ribosome evolved by gene fusion or duplication [40].

In short, the intricate ribosomal architecture positions its substrates in an orientation that promotes peptide bond formation [23, 39, 40] and provides the machinery required for the processivity of this reaction, i.e. for enabling the repetition of peptide bond formation, which results in amino acid polymerization. The current consensus view is that the ribosome contributes positional catalysis to peptide bond formation and provides the path along which A- to P-site translocation occurs, whereas the proximal 2'-hydroxyl of P-site tRNA A76 provides the catalysis [32, 51]. This view answers most of the issues associated with this function, nevertheless further studies are clearly required in order to shed more light on the still unresolved issues, such as the possible involvement of protein L27 in this step [35].

8. The termination step

The hydrolytic cleavage of the ester bond in peptidyl-tRNA during the termination step is also catalyzed by the ribosome. In addition to the participation of ribosomal components, e.g. A2602 ribose [57], peptide release requires auxiliary release factors that recognize the termination codons and promote the P-site peptidyl-tRNA hydrolysis and appear to induce ribosome conformational changes [18]. The disassembly of the ribosome at the end of translation is facilitated in bacteria by the ribosome recycling factor (RRF), in a manner yet to be elucidated. Thus, motions in intersubunit bridges that have been suggested based on a crystal structure of RRF bound to the large ribosomal subunit [28] and to the vacant ribosome [10], were not seen in the crystal structure of T70S in a complex containing a stop codon, a tRNA anticodon in the P-site, tRNAfMet in the E-site and RRF [16]. The mode of E-site tRNA release, its possible involvement in codon–anticodon interactions, and the biological meanings of the different conformations of vacant ribosomes, remain open questions.

9. Nascent protein voyage within the ribosome and its emergence into the cellular environment

Nascent polypeptides progress through their exit tunnel (Figure 3); a universal feature of the large ribosomal subunit that lies adjacent to the PTC [2, 3], and is lined primarily by rRNA with a few r-proteins reaching its walls from its exterior (Figure 3a). This tunnel (~120 Å in length and varying diameter, 10–25 Å) possesses the dynamics required for interacting with the nascent protein. Thus, it seems to play an active role in sequence-specific arrest of nascent chains and in response to cellular signals [68], namely in gating and discriminating, as well as in controlling the operational mode of the translocon at the ER membrane [47]. Tunnel wall elements that appear to sense nascent-peptide specific sequences include, in addition to the rRNA, r-proteins L22 [7, 68] and L4 that form the tunnel's constriction, L23 that in eubacteria extends into the tunnel [25], and a crevice adjacent to the tunnel-wall that can provide space for cotranslational transient folding that was suggested by results obtained by non-crystallographic methods, including FRET measurements [41] and computational analyses [38].

While being translated nascent proteins emerge from their protective exit tunnel into the crowded cellular environment before gaining sufficient length to acquire the final fold. Molecular chaperones support correct folding within the crowded cells. In eubacteria, the first chaperone encountered by the emerging nascent chain, called trigger factor (TF), binds to the translating ribosome at ~1:1 stoichiometry by interacting with ribosomal proteins L23 and L29 [25–27]. Protein L23 belongs to the small group of ribosomal proteins that display significant evolutionary divergence. Whereas its globular domain is conserved [25], only in eubacteria does it possess a sizable elongated loop, which extends from the ribosome exterior all the way into the tunnel walls (Figure 3). At this position, the L23 extended loop can undergo allosteric conformational changes that, in turn, can modulate the shape of the tunnel, which implies trafficking of the nascent protein [25, 26]. Modeling of full-length TF and the signal-recognition particle (SRP) onto the TFa-50S complex suggests simultaneous cohabitation [26] in a fashion that presumably allows screening for hydrophobic signal sequences on the emerging nascent chains [78]. Hence, an interplay between TF, SRP and the trafficked nascent chain while progressing through the tunnel, is plausible.

Based on the structure of unbound TF from *E. coli* [10], the homology between trigger factor from *E. coli* and *D. radiodurans* and analyses of crystal structures of physiologically meaningful complexes of D50S with TF binding domain (called TFa) from the same source [25, 26], it was found that TFa undergoes conformational rearrangements that expose a sizable hydrophobic region (Figure 3), thus acquiring a configuration that is suitable for adherence to hydrophobic patches on the nascent chain. Consistent with dynamic studies [41], it appears that TFa prevents the aggregation of the emerging nascent chain by providing a hydrophobic surface that can transiently mask exposed hydrophobic regions of the elongating polypeptide chains until they become buried in the interior of the mature protein.

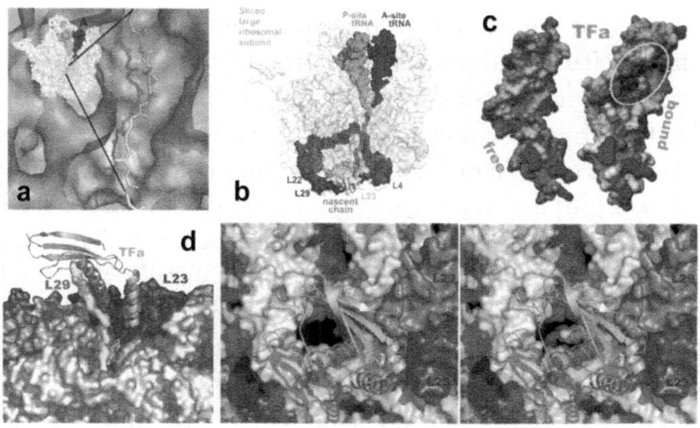

Figure 3. The nascent protein exit tunnel and chaperoning the emerging proteins. (a) The position, the curvature, and the varying diameter of the protein exit tunnel within the large ribosomal subunit are indicated by a modeled polyalanine (yellow). (b) Proteins reaching the tunnel's walls from the large subunit exterior. The tunnel interior is marked by a modeled nascent chain (orange). The large subunit is shown in blue-grey. (c) Conformational differences between free and ribosome bound TFa, based the structure of the homologous complex of TFa and the large ribosomal subunit from D. radiodurans [25] and on the very high level of homology between TF molecules in E. coli and in D. radiodurans. The yellow ellipse delineates the sizable hydrophobic region that becomes exposed upon its binding to the ribosome. The coordinates of E. coli free TFa were taken from [27]. (d) Spacefilling representation of ribosomal RNA (in grey) and r-proteins (in blue, dark red and dark green) at the tunnel opening. TFa is shown as gold ribbons, and a modeled nascent chain as yellow ribbons. Left: the emerging protein (modeled polyalanine) enters the shelter provided by the trigger factor binding domain (TFa). The proteins associated with the trigger factor, L23 and L29, are shown. Note L23 extension reaching the tunnel wall (as shown also in (b)). Middle and Right: a view perpendicular to the view shown in the left, of the tunnel opening. Middle: empty tunnel. Right: A modeled polyalanine chain is emerging from the tunnel. Note that in this crystal structure the tunnel was empty.

10. Strategies taken by antibiotics targeting ribosomes

Despite ribosome conservation many of the antibiotics targeting ribosomes are clinically relevant (reviewed in [56–61, 67–71]). As so far there are no crystals of ribosomes from a pathogenic organisms, structural information is currently obtained from the crystallizable eubacterial ribosomes that have shown to be relevant for determining antibiotic targets of pathogens. These structures have shown that antibiotics targeting ribosomes exploit diverse strategies with common denominators. All antibiotics bind to functionally relevant regions, and each prevents a crucial step in the biosynthetic cycle. These include causing miscoding, minimizing essential functional mobility, inhibiting translation initiation, interfering with tRNA substrate binding at the decoding center, hindering tRNA substrate accommodations at the peptidyl transferase center (PTC), preventing interactions of the ribosomal recycling factor (RRF) and blocking the protein exit tunnel.

Alongside rationalizing many genetic, biochemical and medical observations, the available structures have revealed unexpected inhibitory modes. An example is the exploitation of the ribosomal inherent flexibility for antibiotic synergism [56] and for triggering an induced-fit mechanism by remote interactions that reshape the antibiotic binding pocket [63] and consequently led to therapeutical usefulness of an antibiotic family that binds to conserved functional regions, hence not expected to be clinically relevant.

Among the ribosomal antibiotics, the pleuromutilins are of special interest since they bind to the almost fully conserved PTC, yet they discriminate between eubacterial and mammalian ribosomes. To circumvent the high conservation of the PTC the pleuromutilins exploit the inherent functional mobility of the PTC and trigger a novel induced-fit mechanism that involves a network of remote interactions between flexible PTC nucleotides and less conserved nucleotides residing in the PTC-vicinity. These interactions reshape the PTC contour and trigger its closure on the bound drug [63]. The uniqueness of pleuromutilins mode of binding led to new insights into ribosomal functional flexibility, as it indicated the existence of an allosteric network around the ribosomal active site. Indeed, the value of these findings is far beyond their perspective clinical usage, as they highlight basic issues, such as the possibility of remote reshaping of binding pockets and the ability of ribosome inhibitors to benefit from the ribosome functional flexibility.

The identification of the various modes of action of antibiotics targeting ribosomes and a careful analysis of the ribosomal components comprising the binding pockets confirms that the imperative distinction between eubacterial pathogens and mammalian ribosomes hinges on subtle structural difference within the antibiotic binding pockets [56, 58]. Furthermore, comparisons of the different crystal structures of ribosomal particles in complexes with antibiotics indicate that minute variations in the chemical entities of the antibiotics can lead to significantly different binding modes, and that the mere binding of an antibiotic is not sufficient for therapeutic effectiveness. Thus, the available structures have also helped to identify factors that discriminate between pathogenic bacteria and non-pathogenic eukaryotes, which are of crucial clinical importance, since most ribosomal antibiotics target highly conserved functional sites. Thus, comparisons between the antibiotic binding sites in ribosomes from eubacteria (e.g. from *D. Radiodurans*) and those from the archaeon *H. marismortui*, which shares properties with eukaryotes, highlighted the distinction between binding and inhibitory activity. Specifically, this comparison indicated that the identity of a single nucleotide determines antibiotic binding, whereas proximal stereochemistry governs the antibiotic orientation within the binding pocket [56, 58] and consequently its therapeutic effectiveness. This is in accord with recent mutagenesis studies showing that mutation from guanine to adenine in 25S rRNA at the position equivalent to *E. coli* A2058 does not confer erythromycin sensitivity in *Saccharomyces cerevisae* [71].

The elucidation of common principles of the mode of action of antibiotics targeting the ribosome, combined with variability in binding modes, the revelation of diverse mechanisms acquiring antibiotic resistance, and the discovery that remote interactions can govern induced-fit mechanisms enabling species discrimination even within highly conserved regions, justify expectations for structural based improved properties of existing antibiotics as well as for the development of novel drugs.

11. Concluding remarks

The high resolution structures have shown that all ribosomal tasks are governed by the ribosome architecture and simulated unpredictable expansion in ribosome research, which has resulted in new insights into the translation process. Among the new, less expected, findings are the intricate mode of decoding, the mobility of most of the ribosomal functional features, the symmetrical region, the dynamic properties of the ribosomal tunnel, its interactions with the progressing nascent chains, the possible signaling between the ribosome and cellular components and the way the trigger factor prevents misfolding. In addition, unique structural tools for improving antibiotic targets are now available and key issues associated with the structural bases for antibiotics resistance, synergism, and selectivity can now be addressed. However, despite the extensive research and the immense progress, several key issues are still unresolved, some of which are described above. Thus, it is clear that the future of ribosome research and its applicative aspects hold more scientific excitements.

ACKNOWLEDGMENTS

Thanks are due to all members of the ribosome group at the Weizmann Institute for constant interest. Support was provided by the US National Inst. of Health (GM34360), and the Kimmelman Center for Macromolecular Assemblies. AY holds the Martin and Helen Kimmel Professorial Chair.

References

1 Yonath, A. et al. (1980) Crystallization of the large ribosomal subunit from B. stearothermophilus. *Biochem Int* 1, 315–428
2 Ban, N. et al. (2000) The complete atomic structure of the large ribosomal subunit at 2.4 A resolution. *Science* 289 (5481), 905–920
3 Harms, J. et al. (2001) High resolution structure of the large ribosomal subunit from a mesophilic eubacterium. *Cell* 107 (5), 679–688
4 Schluenzen, F. et al. (2000) Structure of functionally activated small ribosomal subunit at 3.3 angstroms resolution. *Cell* 102 (5), 615–623
5 Wimberly, B.T. et al. (2000) Structure of the 30S ribosomal subunit. *Nature* 407 (6802), 327–339
6 Yusupov, M.M. et al. (2001) Crystal structure of the ribosome at 5.5 A resolution. *Science* 292 (5518), 883–896
7 Yonath, A. (2005) Ribosomal crystallography: peptide bond formation, chaperone assistance and antibiotics activity. *Mol Cells* 20, 1–16
8 Ogle, J.M. and Ramakrishnan, V. (2005) Structural insights into translational fidelity. *Annu Rev Biochem* 74, 129–177
9 Moore, P.B. and Steitz, T.A. (2005) The ribosome revealed. *Trends Biochem Sci* 30 (6), 281–283
10 Schuwirth, B.S. et al. (2005) Structures of the Bacterial Ribosome at 3.5 A Resolution. *Science* 310 (5749), 827–834
11 Selmer, M. et al. (2006) Structure of the 70S Ribosome Complexed with mRNA and tRNA. *Science* 313 (5795), 1935–1942
12 Korostelev, A. et al. (2006) Crystal Structure of a 70S Ribosome-tRNA Complex Reveals Functional Interactions and Rearrangements. *Cell* 126, 1065–1077
13 Yusupova, G. et al. (2006) Structural basis for messenger RNA movement on the ribosome. *Nature* 444 (7117), 391–394
14 Jenner, L. et al. (2007) Messenger RNA conformations in the ribosomal E site revealed by X-ray crystallography. *EMBO Rep* 8 (9), 846–850
15 Jenner, L. et al. (2005) Translational operator of mRNA on the ribosome: how repressor proteins exclude ribosome binding. *Science* 1308 (5718), 120–123
16 Weixlbaumer, A. et al. (2007) Crystal structure of the ribosome recycling factor bound to the ribosome. *Nat Struct Mol Biol* 14 (8), 733–737
17 Pai, R.D. et al. (2008) Structural insights into ribosome recycling factor interactions with the 70S ribosome. *J Mol Biol* 376 (5), 1334–1347
18 Petry, S. et al. (2005) Crystal structures of the ribosome in complex with release factors RF1 and RF2 bound to a cognate stop codon. *Cell* 123 (7), 1255–1266
19 Kaminishi, T. et al. (2007) A snapshot of the 30S ribosomal subunit capturing mRNA via the Shine-Dalgarno interaction. *Structure* 15 (3), 289–297
20 Dunham, C.M. et al. (2007) Structures of tRNAs with an expanded anticodon loop in the decoding center of the 30S ribosomal subunit. *RNA* 13 (6), 817–823
21 Weixlbaumer, A. et al. (2007) Mechanism for expanding the decoding capacity of transfer RNAs by modification of uridines. *Nat Struct Mol Biol* 14 (6), 498–502
22 Nissen, P. et al. (2000) The structural basis of ribosome activity in peptide bond synthesis. *Science* 289 (5481), 920–930
23 Bashan, A. et al. (2003) Structural basis of the ribosomal machinery for peptide bond formation, translocation, and nascent chain progression. *Mol Cell* 11, 91–102
24 Schmeing, T.M. et al. (2005) An induced-fit mechanism to promote peptide bond formation and exclude hydrolysis of peptidyl-tRNA. *Nature* 438 (7067), 520–524

24 Schmeing, T.M. et al. (2005) Structural insights into the roles of water and the 2' hydroxyl of the P Site tRNA in the peptidyl transferase reaction. *Mol Cell* 20 (3), 437–448
25 Baram, D. et al. (2005) Structure of trigger factor binding domain in biologically homologous complex with eubacterial ribosome reveals its chaperone action. *Proc Natl Acad Sci U S A* 102, 12017–12022
26 Schluenzen, F. et al. (2005) The binding mode of the trigger factor on the ribosome: Implications for protein folding and SRP interaction. *Structure (Camb)* 13 (11), 1685–1694
27 Ferbitz, L. et al. (2004) Trigger factor in complex with the ribosome forms a molecular cradle for nascent proteins. *Nature* 431 (7008), 590–596
28 Wilson, D.N. et al. (2005) X-ray crystallography study on ribosome recycling: the mechanism of binding and action of RRF on the 50S ribosomal subunit. *Embo J* 24 (2), 251–260
29 Gregory, S.T. et al. (2005) Mutational Analysis of 16S and 23S rRNA genes of Thermus thermophilus. *J Bacteriol* 187 (14), 4804–4812
30 Bayfield, M.A. et al. (2001) A conformational change in the ribosomal peptidyl transferase center upon active/inactive transition. *Proc Natl Acad Sci U S A* 98 (18), 10096–10101
31 Xiong, L. et al. (2001) pKa of adenine 2451 in the ribosomal peptidyl transferase center remains elusive. *RNA* 7 (10), 1365–1369
32 Weinger, J.S. et al. (2004) Substrate-assisted catalysis of peptide bond formation by the ribosome. *Nat Struct Mol Biol* 11 (11), 1101–1106
33 Youngman, E.M. et al. (2004) The active site of the ribosome is composed of two layers of conserved nucleotides with distinct roles in peptide bond formation and peptide release. *Cell* 117 (5), 589–599
34 Diaconu, M. et al. (2005) Structural basis for the function of the ribosomal L7/12 stalk in factor binding and GTPase activation. *Cell* 121 (7), 991–1004
35 Maguire, B.A. et al. (2005) A protein component at the heart of an RNA machine: the importance of protein l27 for the function of the bacterial ribosome. *Mol Cell* 20 (3), 427–435
36 Beringer, M. et al. (2005) Essential mechanisms in the catalysis of peptide bond formation on the ribosome. *J Biol Chem* 280 (43), 36065–36072
37 Sharma, P.K. et al. (2005) What are the roles of substrate-assisted catalysis and proximity effects in peptide bond formation by the ribosome? *Biochemistry* 44 (30), 11307–11314
38 Ziv, G. et al. (2005) Ribosome exit tunnel can entropically stabilize {alpha}-helices. *Proc Natl Acad Sci U S A* 102: (52), 18956–18961
39 Agmon, I. et al. (2005) Symmetry at the active site of the ribosome: Structural and functional implications. *Biol Chem* 386 (9), 833–844
40 Agmon, I. et al. (2006) On Ribosome Conservation and Evolution. *Isr J Ecol Evol* 52, 359–379
41 Kaiser, C.M. et al. (2006) Real-time observation of trigger factor function on translating ribosomes. *Nature* 444 (7118), 455–460
42 Trobro, S. and Aqvist, J. (2006) Analysis of predictions for the catalytic mechanism of ribosomal peptidyl transfer. *Biochemistry* 45 (23), 7049–7056
43 Bieling, P. et al. (2006) Peptide bond formation does not involve acid-base catalysis by ribosomal residues. *Nat Struct Mol Biol* 13 (5), 424–428
44 Brunelle, J.L. et al. (2006) The interaction between C75 of tRNA and the A loop of the ribosome stimulates peptidyl transferase activity. *RNA* 12 (1), 33–39

45 Sato, N.S. et al. (2006) Comprehensive genetic selection revealed essential bases in the peptidyl-transferase center. *Proc Natl Acad Sci U S A* 103 (42), 15386–15391
46 Gindulyte, A. et al. (2006) The transition state for formation of the peptide bond in the ribosome. *Proc Natl Acad Sci U S A* 103 (36), 13327–13332
47 Woolhead, C.A. et al. (2006) Translation arrest requires two-way communication between a nascent polypeptide and the ribosome. *Mol Cell* 22 (5), 587–598
48 Uemura, S. et al. (2007) Peptide bond formation destabilizes Shine-Dalgarno interaction on the ribosome. *Nature* 446 (7134), 454–457
49 Helgstrand, M. et al. (2007) The ribosomal stalk binds to translation factors IF2, EF-Tu, EF-G and RF3 via a conserved region of the L12 C-terminal domain. *J Mol Biol* 365 (2), 468–479
50 Rodnina, M.V. et al. (2007) How ribosomes make peptide bonds. *Trends Biochem Sci* 32 (1), 20–26
51 Weinger, J.S. and Strobel, S.A. (2007) Exploring the mechanism of protein synthesis with modified substrates and novel intermediate mimics. *Blood Cells Mol Dis* 38 (2), 110–116
52 Hobbie, S.N. et al. (2007) Engineering the rRNA decoding site of eukaryotic cytosolic ribosomes in bacteria. *Nucleic Acids Res* 35 (18), 6086–6093
53 Youngman, E.M. et al. (2007) Stop codon recognition by release factors induces structural rearrangement of the ribosomal decoding center that is productive for peptide release. *Mol Cell* 28 (4), 533–543
54 Cho, P.F. et al. (2005) A new paradigm for translational control: inhibition via 5'-3' mRNA tethering by Bicoid and the eIF4E cognate 4EHP. *Cell* 121 (3), 411–423
55 Andersen, C.B. et al. (2006) Structure of eEF3 and the mechanism of transfer RNA release from the E-site. *Nature* 443 (7112), 663–668
56 Yonath, A. and Bashan, A. (2004) Ribosomal crystallography: Initiation, peptide bond formation, and amino acid polymerization are hampered by antibiotics. *Annu Rev Microbiol* 58, 233–251
57 Polacek, N. and Mankin, A.S. (2005) The ribosomal peptidyl transferase center: structure, function, evolution, inhibition. *Crit Rev Biochem Mol Biol* 40 (5), 285–311
58 Yonath, A. (2005) Antibiotics targeting ribosomes: resistance, selectivity, synergism, and cellular regulation. *Annu Rev Biochem* 74, 649–679
59 Tenson, T. and Mankin, A. (2006) Antibiotics and the ribosome. *Mol Microbiol* 59 (6), 1664–1677
60 Bottger, E.C. (2007) Antimicrobial agents targeting the ribosome: the issue of selectivity and toxicity - lessons to be learned. *Cell Mol Life Sci* 64 (7–8), 791–795
61 Schluenzen, F. et al. (2006) The antibiotic kasugamycin mimics mRNA nucleotides to destabilize tRNA binding and inhibit canonical translation initiation. *Nat Struct Mol Biol* 13 (10), 871–878
62 Schuwirth, B.S. et al. (2006) Structural analysis of kasugamycin inhibition of translation. *Nat Struct Mol Biol* 13 (10), 879–886
63 Davidovich, C. et al. (2007) Induced-fit tightens pleuromutilins binding to ribosomes and remote interactions enable their selectivity. *Proc Natl Acad Sci U S A* 104 (11), 4291–4296
64 Pyetan, E. et al. (2007) Chemical parameters influencing fine-tuning in the binding of macrolide antibiotics to the ribosomal tunnel. *Pure Appl Chem* 79 (6), 955–968
65 Borovinskaya, M.A. et al. (2007) Structural basis for aminoglycoside inhibition of bacterial ribosome recycling. *Nat Struct Mol Biol* 14 (8), 727–732

66 Schroeder, S.J. et al. (2007) The structures of antibiotics bound to the E Site region of the 50 S ribosomal subunit of Haloarcula marismortui: 13-Deoxytedanolide and Girodazole. *J Mol Biol* 367 (5), 1471–1479
67 Hobbie, S.N. et al. (2008) Mitochondrial deafness alleles confer misreading of the genetic code. *Proc Natl Acad Sci U S A* 105 (9), 3244–3249
68 Berisio, R. et al. (2003) Structural insight into the role of the ribosomal tunnel in cellular regulation. *Nat Struct Biol* 10 (5), 366–370
69 Pfister, P. et al. (2005) 23S rRNA base pair 2057-2611 determines ketolide susceptibility and fitness cost of the macrolide resistance mutation 2058A–>G. *Proc Natl Acad Sci U S A* 102 (14), 5180–5185
70 Tu, D. et al. (2005) Structures of MLSBK antibiotics bound to mutated large ribosomal subunits provide a structural explanation for resistance. *Cell* 121, 257–270
71 Bommakanti, A.S. et al. (2008) Mutation from guanine to adenine in 25S rRNA at the position equivalent to E. coli A2058 does not confer erythromycin sensitivity in Sacchromyces cerevisae. *RNA* 14 (3), 460–464
72 Frank, J. et al. (2005) The role of tRNA as a molecular spring in decoding, accommodation, and peptidyl transfer. *FEBS Lett* 579 (4), 959–962
73 Konevega, A.L. et al. (2007) Spontaneous reverse movement of mRNA-bound tRNA through the ribosome. *Nat Struct Mol Biol* 14 (4), 318–324
74 Frank, J. and Agrawal, R.K. (2000) A ratchet-like inter-subunit reorganization of the ribosome during translocation. *Nature* 406 (6793), 318–322
75 Wittmann, H.G. (1982) Structure and evolution of ribosomes. *Proc R Soc Lond B Biol Si*, 216, 117–135
76 Diedrich, G. et al. (2000) Ribosomal protein L2 is involved in the association of the ribosomal subunits, tRNA binding to A and P sites and peptidyl transfer. *Embo J* 19 (19), 5241–5250
77 Sobolevsky, Y. and Trifonov, E.N. (2005) Conserved sequences of prokaryotic proteomes and their compositional age. *J Mol Evol* 61 (5), 591–596
78 Schaffitzel, C. et al. (2006) Structure of the E. coli signal recognition particle bound to a translating ribosome. *Nature* 444 (7118), 503–506

GPSR Compliance

The European Union's (EU) General Product Safety Regulation (GPSR) is a set of rules that requires consumer products to be safe and our obligations to ensure this.

If you have any concerns about our products, you can contact us on ProductSafety@springernature.com

In case Publisher is established outside the EU, the EU authorized representative is:

Springer Nature Customer Service Center GmbH
Europaplatz 3
69115 Heidelberg, Germany

Batch number: 10584469

Printed by Printforce, the Netherlands